# Customization
## WHY LIVE IN THE BOX?

# Table of Contents

*Photo by Jared Burks.*

# Minifigure Customization2: Why Live In A Box?

Author: Jared Burks

Layout Artist: Joe Meno

Contributiing Photographers: Kenneth Goh Kay Boon, Michael "Xero" Marzilli. Scott Berg, Matt Sailors, Noah Glenn, Jeff Lee, Sebastian Sand, Glen Wadleigh, Iain Heath, Mark Parker, Jimmy "THE BOXMAN" Chavez, Michael Inglis, Rob Hendrix

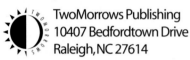

TwoMorrows Publishing
10407 Bedfordtown Drive
Raleigh, NC 27614
www.twomorrows.com • e-mail: twomorrow@aol.com
First Printing: November 2013 • Printed in China
ISBN-13: 978-1-60549-052-6

## Dedication & Acknowledgement

To my wife who supports me daily and my daughter whose curiosity and interest in "what daddy is doing" is a renewing source of energy and motivation to create new things. Watching Branwen, at 4 years old, pick up a decal and apply it correctly was simple, yet amazing. Giving her no direct instruction beyond get this wet and move it there, she figured out the whole process near instantaneously. I hope this book serves as a guide to what is possible and others learn from doing, as did my daughter. Her achievement brought on instant pride and desire to show others her creation and how it was achieved. This made a MiniMaker's Faire all the more memorable. Share not only your work, but your process. Why live in a box? Amber and Branwen, thank you to you both for being part of my life.

I would also like to thank a few friends who have helped me and shared in the insanity; Chris, Mark, Michael, and Matt, many thanks.

I would also like to thank Joe and John for giving me the opportunity to write these books and the article series for BrickJournal.

In 1932, Ole Kirk Kristiansen began making wooden LEGO toys. Godtfred Kirk Christiansen (Ole's son) submitted the patent for the LEGO brick in 1958. 1974 saw the birth of the LEGO figure and in 1978, the minifigure, as we now know it, was created. By this time, Ole's grandson, Kjeld Kirk Kristiansen, was running the LEGO Group. Since then, the population of the minifigure has eclipsed 4 billion, making it the world's largest population group. Since 1990, LEGO has also allowed this tiny figure to take on a wider range of expressions: typically on the reversible double-sided heads we get the bonus expressions of brave, bold, or frightened. (For more on this trend, you can read the following: *http://bartneck.de/publications/2013/agentsWithFaces/bartneckLEGOAgent.pdf*.) I, for one, am grateful for this diversity in expression.

In 2010, the biggest change to LEGO since 1978 occurred: the debut of the Collectible Minifigure Series. It is this series that altered the double-sided expression heads, giving a wider range of expressions on a single side of the minifigure head. This was expanded with the various superhero licenses with DC and Marvel. The Collectible Minifigure Series also started many a personal hunt for that one special figure. This hunt has been recently intensified with the addition of very rare Mr. Gold in Collectible Minifigure Series 10.

Initially the collectible minifigure packages featured bar codes, making identification easy, however LEGO quickly rectified this and replaced bar codes with small round indentions referred to as bump codes. It didn't take LEGO enthusiasts long to crack these bump codes either. Today the LEGO customizer typically blindly feels the package for that special minifigure element that sets the figure apart from all others. As a result, all of the series minifigures have supplied customizers with a new and diverse palette of parts, accessories, and hairpieces, which are all highly desired. This influx of parts, especially hairpieces, has been exactly what customizers have needed. I believe it was critical that LEGO tackle a project of this nature as the aftermarket for custom parts continues to grow with new businesses opening all the time. As such, LEGO is really starting to understand the power of the minifigure, which is evident when one examines the new LEGO sets, license properties that LEGO has secured, and the swag merchandise that feature a unique figure (books, movies, video games, etc.). These unique figures can even be virtual; several video games have offered exclusive in-game characters with pre-order. I believe the reason behind this power is due to the fact that the level of detail in the "standard" figure has dramatically improved. LEGO is finally making accessories that resemble their actual items,

*The first minifigure lineup.*

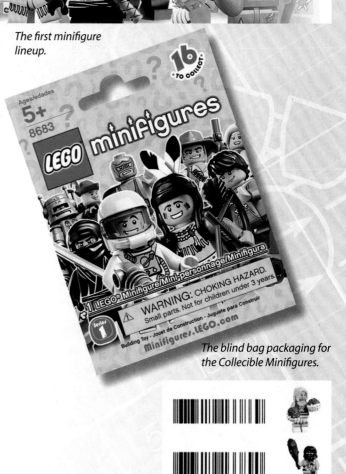

*The blind bag packaging for the Collectible Minifigures.*

*Right: Barcodes for Seris 1 minifigures.*

*Below: Bumpcodes for a later minifigure series.*

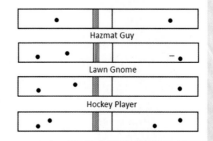

Hazmat Guy

Lawn Gnome

Hockey Player

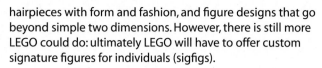

hairpieces with form and fashion, and figure designs that go beyond simple two dimensions. However, there is still more LEGO could do: ultimately LEGO will have to offer custom signature figures for individuals (sigfigs).

Recently, the *Minifigures Character Encyclopedia* has been released by DK Publishing. This book features tons of behind- the-scenes information about how and why the figures feature the details that they do. It is this backstory that actually makes the figures better. Why you might ask? Well because the details have a reason for being there and they are artist-specific. Specifically, let's examine the LEGO cheerleaders from S1 and S8, which features an M and A (respectively) on their sweaters. Only when knowing that the LEGO Vice President in charge of the Minifigures line is Matthew Ashton does these letters take meaning. There are several such details added throughout the line for the various artists and creators behind the series. I believe that the line excels because the artists/creators are so invested because these figures carry their monikers.

The Collectible Minifigure series has also been following a typical LEGO idea: themed characters. When the series is examined globally almost every character fits into an existing LEGO theme, allowing characters across the series to function together and with existing system sets. Examining the themes LEGO has created for the system sets gives a glimpse into the Collectible Minifigure themes.

Where LEGO has failed with the Collectible Minifigure series is the development of extensively new themes. By creating figures that are outside of the existing LEGO themes it would allow LEGO to test the market for newer themes in a less expensive way. The only completely new themes have been the Party Animals and the ancient Greek/Romans/Mythology. It would be nice if LEGO, in the architecture line give us an ancient Coliseum for our Gladiators and Roman's to skirmish! Have you thought about the party room with the lampshade hat to go along with the Party Animals?

Once you group the figures, all of the above are possible and you quickly see that LEGO has been creating figures in larger groups. LEGO has started to think a bit about the theme sets and in later iterations of the series started to supply "friends" for the sole figures.

*Photos of minifigures by WhiteFang (Kenneth Goh Kay Boon).*

## Minifigure Themes

| Theme | Counts | Theme | Counts | Theme | Counts | Theme | Counts |
|---|---|---|---|---|---|---|---|
| City | 37 | Castle | 7 | Police | 3 | Pirate | 2 |
| Sports | 21 | Old West | 7 | Revolutionary | 3 | Prehistoric | 2 |
| Fantasy | 16 | Musician | 6 | Skater | 3 | Scottish | 2 |
| Performer | 15 | Roman | 5 | Surfer | 3 | Skier | 2 |
| Classic Monster | 14 | Archaeology | 4 | Actor | 2 | Snowboarder | 2 |
| Circus | 10 | Dance | 4 | Cowboy | 2 | Tarzan | 2 |
| X-Games | 10 | Robot | 4 | Diver | 2 | Tennis | 2 |
| Space | 9 | Asian | 3 | European | 2 | Viking | 2 |
| Water | 9 | Aztec | 3 | Hospital | 2 | Workout | 2 |
| Greek | 9 | Egyptian | 3 | Military | 2 | Cheerleader | 2 |
| Party Animals | 8 | Indian | 3 | Motorcycle | 2 | | |

In series 10 of the Collectible Minifigures, LEGO has taken a page out of Willie Wonka's book and added a 17th figure. This 17th figure is a gold ultra-rare chase figure named "Mr. Gold" (5000 inserted into the series). LEGO has also altered the distribution of the figures. Until series 9, they were widely available in everything from the LEGO store to grocery and convenience stores. LEGO has reduced this availability in an effort to draw consumers into stores that carry larger selections of their product line. With the change in distribution policy and the creation of the magical 17th figure I believe LEGO will achieve their desired affect and help boost sales across their product line. However, it also exposes consumers to more options as several other companies have been hot on their heels creating their very own collectible series. Ultimately, I prefer the presence of LEGO in more stores. This exposes more children to LEGO and while the initial purchase might be limited to a single figure, the child that receives the figure will ultimately need to build that figure a vehicle, structure, or environment.

One of the best reviewers of the Collectible Minifigure Series is WhiteFang 'Kenneth Goh Kay Boon.' Keep an eye on his Flickr gallery (*http://www.flickr.com/photos/whitefang_eurobricks/*) to keep up to date as newer series are released. Also check out Eurobricks, where he posts his reviews: *http://www.eurobricks. com/forum/index.php?showtopic=80676.*

While not related to the Collectible series there is an ultimate guide on LEGO minifigure taxonomy, check out the "Taxonomy for LEGO Minifigures" by Christoph Bartneck - See more at: *http://www.bartneck.de/2010/12/17/taxonomy-for-LEGO-minifigures/#sthash.XteF830J.dpuf*

*Mr. Gold.*

*Here are just some of the great items LEGO has given customizers: hockey sticks, skates, minifigure trophies, medals, gavels and paintball equipment! Photos by Michael "Xero" Marzilli.*

I continually ask myself this question. The reason I ask is to push the limits of the answer; the only way to do this is to think about the limits created in the hobby. So I ask, "What is customizing?" Let's think about what I am really asking in this question. I believe this question is twofold. The basic question I am asking is, *"What is art?"* The secondary question is, *"What is minifigure customization to me?"*

After researching and thinking about the question, *"What is art?"* for some time I found many answers. I am not sure which answer is correct, I believe it is likely different for different people as art is personal. However, many of the answers I come up with and I find through research all share the same foundation. The basic answer I can give for art is that it contains both form and content. The definition of art when searched on Google is:

### Art /ärt/, *Noun*

> **1.** The expression or application of human creative skill and imagination, typically in a visual form such as painting or sculpture..."the art of the Renaissance"
>
> **2.** Works produced by such skill and imagination.

*Art contains **form** and **content** created by **imagination** through the use of **skill**.*

While this simple sentence defines art and how it is created, it is a very complex thought. It also speaks to what is customizing. Since the purpose of this book is to help the readers develop the skills to create the form and content these are all critical ideas to examine.

### Form, *Noun*

**1.** The elements of art

**2.** The principles of design

**3.** The actual, physical materials that the artist has used.

The form is basically what you want the finished figure to look like. Do you want to craft custom figures to look at home next to official LEGO Minifigures, or do you want to make them look more life-like, or do you want them to look entirely different? These simple choices will influence you a great deal when you sit down to create your first custom minifigure. It will influence the materials you use and the design principles employed.

### Content, *Noun*

1. What the artist meant to portray,

2. What the artist actually *did* portray and

3. How we react, as individuals, to both the intended and actual messages.

Content is a more complex concept. In the first book, I compared a custom figure to a sentence, which requires a complete thought. Content is that complete thought. When you are creating a custom minifigure, the content of your figure needs a complete thought. What are you trying to portray in the figure? Is he a tough guy? Are you trying to

*Jared's minifigure of the T-Mobile Girl.*

make him a funny tough guy? How do you want people to react to your figure? Do you want them to think it is gruesome, make them laugh, smile? All good figures bring a reaction out of people, what is the reaction you want? This is content. Minifigure expressions are critical in portraying content, which is what LEGO was missing for ages. The Collectible Minifigure series has done a better job, but typically LEGO wants figures to portray humor, happy, or scary. They need to learn that there is other content that figures can portray, this is where the customizer can improve on what LEGO has given.

This book will show the forms and materials that I have used to create custom figures. It will show the content that I have used to portray a wide range of thoughts. I can demonstrate these to you, but you will have to find your own, which is what makes the figure you create your own work of art. Let's re-examine the statement above, "*Art contains **form** and **content** created by **imagination** through the use of **skills**.*" The imagination required also comes from the artist. I can let you peek into my imagination to see how I have developed new ideas for customizing and found new inspirations, but tapping into your own imagination is critical. Skills, are developed and honed, and this takes time and patience. I will encourage you to not give up, as becoming good at any skill takes practice and failures. Only after failures will skill develop and lead to rewards.

Examining the secondary question, "*What is minifigure customization to me?*" is a bit more difficult to define. The information above gives enough background to answer this question. I believe I will let my work displayed here in this book and online, allow you, the reader, decide on your own. I don't want to influence your answer as ultimately this will influence your art and thus your creations.

Through my research for this chapter I found a quote from a famous Italian artist, Michelangelo Pistoletto that I believe captures the reasons we make the figures we do. I will end this section with his words on the subject:

"*Above all, artists must not be only in art galleries or museums — they must be present in all possible activities. The artist must be the sponsor of thought in whatever endeavor people take on, at every level.*"

The Tick.

Clutch Powers .

Papa Smurf.

Colonel Clown, Clown Commandos, *Big Red Boot Entertainment, LLC.*

Zoe the Zookeeper, Mascot for Houston Zoo (Houston, TX).

*Roman Legion by Scott Berg utilizing variation of parts to create officers and troops. Note use of minifigure trophy as flag stand.*

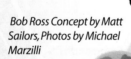

*Gladiators by Scott Berg utilizing various Collectible Minifigure parts.*

*Space Marines by Michael Marzilli. A touch of paint adds quite a bit of detail to create the officer.*

With the Collectible Series a new area of customizing has emerged: the customization of the collectible figures. Customizers have taken the concepts offered by the LEGO Group and finished their thoughts, so to speak, by taking these official figures and adding that detail that the LEGO Group left out. In this chapter, I will point out some of the excellent custom figures that people have produced from the official line. This type of customization more closely reflects what many brick customizers do with official LEGO sets. Sets are beefed up to improve their appearance, structure, and overall level of detail (sometimes this is done only to improve their swoosh-ability). These same three improvements are the same that minifigure customizers are bringing to the Collectible Minifigure Series.

The collectible figure has created a unique niche in the LEGO world. Many strive to create environments for these lovely collectible Minifigures, which has created one of the offshoots of customizing. This blends the custom, or this case the official figure with the world of custom figures. This blending happens by three main methods: customized versions & purist customization, customized/modified parts and customized environment incorporation.

The first method is quite simple. LEGO has supplied customizers with a single figure in large quantities. However we want a team, group, or unit of these figures, this means we must alter them slightly to make each appear unique and part of the whole. It is critical that we have a concept in mind for these figures. Something like the Roman phalanx might need great similarity among the ranks, but the officers need to be easily recognized.

While mixing and matching parts is an easy way to create multiple members of a group, this is the technique behind purist customization, which was extensively discussed in the first book. Essentially the customizer utilizes official LEGO elements to create new figures.

*Killer Bees. (John Belushi & Dan Aykroyd) Concept by Matt Sailors, Photos by Michael Marzilli.*

*Magician? Concept by Matt Sailors, Photos by Michael Marzilli*

*Bob Ross Concept by Matt Sailors, Photos by Michael Marzilli*

The second method is likely the most popular. It is simply adding details or customizing/modifying the parts to make the figure that much better. The best example of this modification is the classic deep sea diver by Noah Glenn. By simply adding the tarnished paint affect to the brass helmet and boots it takes this figure to the next level. However, Glenn went further by adding the dents and dings to the helmet and giving the diver a spear.

Another example of this type of customization is shown in the highly detailed gladiator. This is an area LEGO does not go. While LEGO will build warriors, it will not show battle damage or, more specifically, blood. While the gladiator at the right does have a bit of splatter, it is the sword and shield damage is what really speaks to the battle-worn nature of this character. Clearly he is still ready for battle as his helmet and protective mask are on and his sword and shield are at the ready.

The third improvement is customized environment incorporation. This ranges from vignette building around the collectible figure to simply adding the figure to a small stand, as is the case with Lady Liberty seen on the right.

With the few tweaks, these various customizers have applied to these figures they have completed the sentence by completing the thought that LEGO had started in their creation. To me, this what LEGO is supposed to do: allow the user to create something new. So in my mind, the Collectible Minifigure series is a start, not an end to allow users to make so many more characters and populate their LEGO-verse if you only dare to think out of the box (or blind bag in this case).

*The Diver by Noah Glenn really completes the LEGO official figure by damaging and adding paint to the parts.*

*The Gladiator by Noah Glenn adds the realism to this figure by converting to flesh tone and adding some battle damage.*

*LEGO Statue of Liberty by Sebastian "q_159" Sand. By creating the Liberty Island pedestal and Fort Wood the figure is much more complete.*

*Vignettes by Jeff Lee.*

*Stunt Ped by Michael Marzilli. By placing the stunt rider on the Vespa moped the context of the stunt man is completely changed and made much more comical.*

Without the right tools it is hard to complete any tasks. Therefore this chapter is dedicated to the tools that I have found useful. Please note sources are typically referenced for the US, where I am located. For simplicity, I am grouping the tools by customization technique that they are primarily used, however many tools will be helpful to more than one technique. The listed tools are by no means absolutes, there is always other options. If the ones I suggest don't work for you, find tools that do. This is merely a guide. Clearly the one group of parts you can't avoid is minifigure parts, which will not be covered here, but you can check Bricklink and LEGO Pick-a-Brick. This chapter will not cover the tool use, merely what is available, where to find them, and what skill set they apply to. The point of this chapter is to point you in the right direction early and hopefully save you a few dollars on the tools you do purchase.

## Decaling

The foundation of minifigure customization lies in custom decals. Waterslide decal film is used to affix a new design to the minifigure parts. This part of the hobby requires a few items: vector art software, waterslide decal film, printers, decaling solutions, and application tools.

There are several vector art programs out there including CorelDRAW (my favorite), Adobe Illustrator, Draw Plus, and Inkscape. The last two are mentioned specifically because they are free, which are great alternatives to the high-priced commercial options. Another option to consider is a slightly older version of one of the commercial options. Having the latest version is not a must for creating great art; I still use CorelDraw X4 even though X6 is the current option. These older versions can be purchased for very little money on eBay or other discount sites.

Waterslide decal film can be found almost everywhere these days (hobby and art shops) including Wal-Mart. Typically decal film is $1-2 per sheet. You can pick it up locally to avoid shipping fees. While these brands aren't the best quality, they are not a bad to start with as they are generally a bit thicker and thus easy to apply. The most commonly found film is from Testors. If you can't locate this option, Micro-Mark offers an excellent film. Testors' film is strictly for inkjet printers, whereas laser and inkjet options are available from Micro-Mark. You must choose film that is designed to use with your printer (laser or inkjet). Please review the section on decal printing (page 22), especially if you choose an inkjet option as the decals MUST be sealed prior to dipping them in water.

The decal film type brings us to the next piece of equipment required: a printer. If you hunt around, you can find a real bargain on a printer that is if you don't already have one. If you don't mind internet purchasing, check Spoofee.com or other bargain-finding sites for a few days and I am sure a deal will pop up. I recommend Epson printers due to their inks; check the decal section for details.

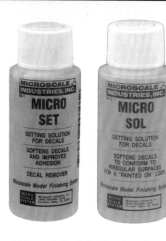

*Microscale Set & Sol decal application solutions.*

Decal application tools are the next items required. Decal setting and softening solutions are critical to advanced application techniques and highly recommended. These solutions can be purchased at most hobby/model stores. The typical brands are Model Masters and Microscale. The Model Master variety can be found at Micro-Mark, Microscale can also be found online, but is typically easy to find locally in model train/model shops (and some Hobby Lobbys). An alternative to these two that works reasonably well is diluted white vinegar (2 drops of water to 1 drop of vinegar). To apply these items you will need a small brush (I prefer inexpensive nylon brushes) which can commonly be found at dollar stores. I prefer nylon because they seem to last a bit longer. Wood stick cotton swabs and tweezers are also very helpful and can be purchased economically at a pharmacy. Decal sealant or clear paint, is also needed; spray cans offer the easiest application. Model Masters has an excellent clear lacquer series. Semi-gloss is pictured; however, I prefer the ultra-clear gloss. The ultra-clear gloss gives a high shine and depending on the figure might not be appropriate. There is an alternative: the $3-5 clear Krylon at Wal-Mart works well. Before spraying, ALWAYS make sure the nozzle is clean so it won't splatter when you spray your decals or figures (**TIP:** to clean a nozzle invert the can and spray it upside down for a few seconds to remove any possible splatter). **QUICK TIP:** Verify you have no silvering below the decal. Once the decal is sealed to the figure, the cause of the silvering (the moisture or air pockets), is trapped and could destroy your decals and the overcoat will magnify the silvering affect.

*Model Masters, Cotton swabs and paintbrush, for decal work.*

| Decaling | | |
|---|---|---|
| **Software** | | |
| Draw Plus | www. freeserifsoftware. com/software/drawplus/ | Free |
| Inkscape | www. inkscape. org | Free |
| **Printer** | | |
| Printer | www. spoofee. com for deals | ~$50 |
| **Decal Film** | | |
| Testors Custom Decal | www. testors. com/product/0/9198/_/Custom_Decal_System | ~$10 |
| Micro-Mark Decal Film | www. micromark. com/decalling. html | ~$10 |
| **Total Cost** | | **~$60** |

# Parts Modification/Creation & Color Alteration: Knives, Mod Tools, Sandpapers, Paints, etc.

Modification tools can vary widely from one project to another, but the basics are hobby knives (X-acto®) and rotary (Dremel®) tools. If you buy name-brand versions of these tools prepare to spend quite a bit of money, however very nice alternatives can be found at Harbor Freight for very little money. Harbor Freight is a US brick and mortar with an online store, so if one isn't near you buy online. If you choose to go into the store make sure to print the item details from the online store as the brick and mortar prices are slightly higher, but they will price match their web store. Remember that knives are sharp and can cut you as easily as your project, therefore I recommend a Kevlar glove, and these are found online at wood carving stores or also at Harbor Freight.

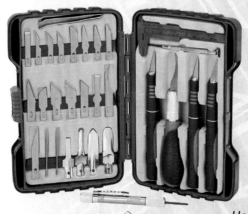

*Hobby knife set.*

Please know that many alterations can be achieved with sandpaper, which is safer for your fingers than a knife. A local home improvement store will have sandpaper, but I doubt you will find any paper of grit 800 or higher. To polish plastic back to a high shine you really need a very high grit paper (~12,000). Micro-mesh makes professional grade cloth backed sandpaper that will last for quite some time. They are expensive compared to the home improvement store option, but in this case I completely believe it is worth the investment. Micro-mesh makes a kit for wood turners to make writing pens and there are resellers on eBay that have these cheaper than anywhere else I have found. The turners' kit contains 9 sheets of 3 x 6 inch sandpaper with gradually increasing grit increments from 1500 – 12,000. Using these in series will leave sub-micron (VERY TINY) scratches that are only visible to a microscope, thus leaving your project with a high shine. TIP: These papers can also be used in sculpting or for the removal of designs from minifigure parts.

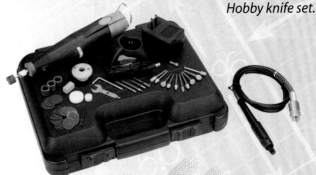

*Rotary tool set.*

## Paints and Dyes

Of all the tools needed, paints are the easiest to find. You can pick up Testors hobby paints cheaply at most stores. Look for the primary color bulk paint packs, these typically run $5-10 and have 8-15 paints of various color in them. These paint packs are a very good value for your money. With a primary color pack, you can mix and make most any color. The hobby uses small items that don't require much paint, so mixing your own paint with a few drops from a bottle works great. Sometimes these packages come with brushes, sometimes they don't. Recall from the previous section that dollar stores commonly offer assorted brush packages, so one

*Kevlar glove.*

*Micro-mesh.*

11

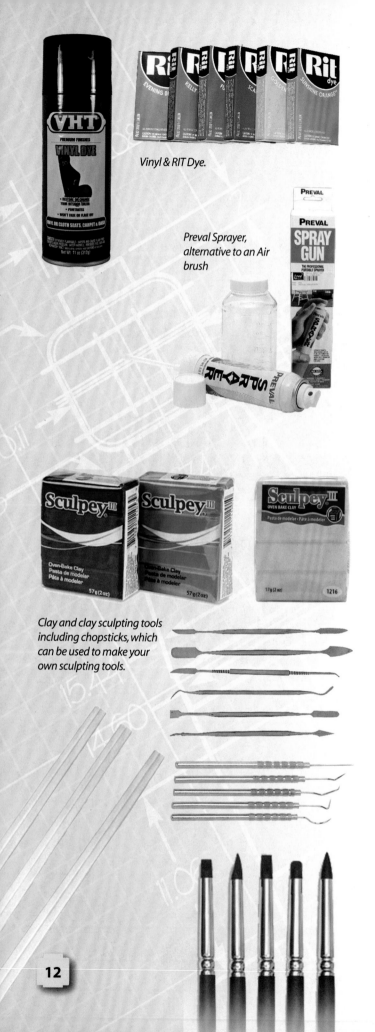

*Vinyl & RIT Dye.*

*Preval Sprayer, alternative to an Air brush*

*Clay and clay sculpting tools including chopsticks, which can be used to make your own sculpting tools.*

purchase can have two functions (painting and decaling). Even if most brushes in the package are not usable, the few that are will be worth the price, just check an art store's pricing for one or two brushes. I prefer nylon brushes for their durability and they are easily identified as they are commonly white or brightly colored bristled brushes.

There are alternatives to painting such as vinyl or fabric dyes. These are much more permanent alterations to the parts and typically cannot be removed by scratching the part as they absorb into the plastic. Vinyl dyes can be found at automotive stores, the drawback is it is typically only available in limited colors. Just be sure you are purchasing a vinyl dye and not a vinyl colorant. Refer to the color alteration section for the difference (page 32). Fabric dyes are available most anywhere including Wal-Mart. The powder versions seem to work better, so search these out instead of the liquids. **TIP:** If using the powdered dyes heat the dye to drive the powder into solution better, continue to heat as you use it to dye the part.

An airbrush is a nice tool to have, but not always cheap. The low-cost option is about $30, but these are really only good for broad coverage and don't always meet every need. Investigate this tool heavily and only purchase when you are sure you have a need. A much cheaper alternative to an air brush is a Preval paint-sprayer. This system has a container you can add any paint to and turn it into a spray paint.

## Sculpting: Clays, Waxes, and Tools

Packages of polymer sculpting clay are very economical and can be found most anywhere. You can work with the clay with homemade tools: paperclips, tongue depressors or Popsicle sticks, and anything else small. The best option, I have recently discovered from a MAKE video featuring action figure creator Scott Hensey, is to use chopsticks tools to sculpt. Cut the chopsticks into any shape you need to make your own sculpting tools—these are an incredible inexpensive option. If you are dead set on commercial tools an economical option is available from Harbor Freight. This is what I commonly use as I found them before the chopsticks tip. There are more expensive options at art stores that have rubber tips or made from surgical stainless steel that are quite nice, so ask yourself how much sculpting you are going to do. Most often the lower-priced or even better free tools (left over from Chinese takeout) are more than enough.

Baking polymer sculpting clay uses your home oven (no expense) or you can cure it in near boiling water. Practice clay sculpting BEFORE getting into molding and casting as this part of the hobby can get quite expensive. Not every piece needs to be molded.

Following the concept of making higher quality/more advanced parts after a clay option is created a wax version can be made. Wax can hold a higher level of detail and gives a rigidity to the part that allows the user to work more easily on the whole part. Wax sculpting typically starts with a molded clay part, which is then cast in wax. I got my recipe for wax from Scott Hensey— the ingredients create a very firm wax.

• 240 grams of Carnauba

• 240 grams of Candelilla

• 60 grams of Bee wax

• 1350 grams of Paraffin wax

• 1500 grams of Talc

• Add Crayons to pigment

Wax can be found online via Amazon or some online cosmetic companies. This is for the very advanced user. You will need a melting pot and a thermometer to use the wax and possibly even a waxer, more details can be found in Chapter 7 as this is beyond the average user.

## Molding and Casting (Kits) & Vacuum Forming

Molding and casting can get expensive quickly and are more technically demanding than any other part of the hobby, especially if you pressure cast. If you choose to go this route, check what is available locally. I recommend starting with a kit that contains both rubber and resin: both Smooth-On and Micro-Mark offer these types of kits. I don't recommend Alumilite as you get half the volume of rubber when compared to other options for the same price. As this is the most expensive and complex off-shoot of the hobby, just as with the airbrush, weigh your needs before purchasing. Details on pressure casting will appear in that chapter due to the specifics.

Vacuum forming is another form of molding and creating new parts in plastic. This method uses a vacuum to pull a thin sheet of heated plastic against a mold to create a new part. It is typically used in creating packaging for items, but many toys and other useful parts are created by this same method. A good used dental vacuum former works well for this task, which can be found on eBay. Thermoplastic sheets (PETG) can be found at your local plastics store or via the internet. Price varies based on thickness.

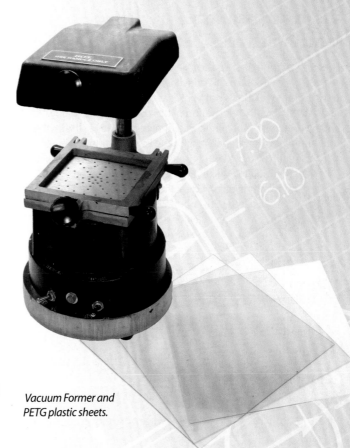

*Vacuum Former and PETG plastic sheets.*

## Parts Modification/Creation & Color Alteration

| General | | |
|---|---|---|
| Hobby Knives | www.harborfreight.com/33-piece-deluxe-hobby-knife-set-96551.html | $13 |
| Hobby Knives | www.harborfreight.com/13-piece-precision-knife-set-32099.html | $3 |
| Rotary Tool | www.harborfreight.com/catalogsearch/result?q=rotary+tool | $8-30 |
| Kevlar Glove | www.harborfreight.com/police-style-kevlar-gloves-large-95568.html | $8 |
| Micro-mesh | micro-surface.com/index.php?main_page=product_info&cPath=273_188_189&products_id=53 | ~$15 |
| **Paint** | | |
| Testors Paints | www.testors.com | $10 |
| Dyes | Vinyl or Fabric - Auto part store or Wal-Mart | ~$5 |
| Preval Sprayer | www.prevalspraygun.com | ~$8 |
| **Sculpting** | | |
| Sculpey III Clay | www.sculpey.com | ~$2 |
| Wax | www.amazon.com | ~$60 |
| Carving Tools | http://www.harborfreight.com/6-piece-stainless-steel-carving-set-34152.html | $5 |
| Clay Shapers | www.texasart.com/g11/Colour-Shaper-Modeling-Tools.htm | $6 |
| **Molding and Casting** | | |
| Smooth-On Kit | www.smooth-on.com/Getting-Started-Po/c4_1217/index.html | $50 |
| Micro-Mark Kit | www.micromark.com/COMPLETE-RESIN-CASTING-STARTER-SET,8174.html | $90 |
| Vacuum Former | www.ebay.com | $300+ |
| **Total Cost** | | **$120** |

## Cloth Accessories:
## Wad Punch and Cloth

For cloth, outside of the material and a pair of scissors, there is really only one tool, a multi-punch leather tool. A leather multi-punch works well, but is not as nice as what Mark "MMCB" Parker uses. Mark drives the innovation of this portion of the hobby and he prefers a set of metric wad punches. However, the leather multi-punch will give good results and is significantly cheaper. Harbor Freight also carries this item as well. Finding a metric set of punches is quite difficult, an English version can be used, however it doesn't create holes to the exact dimensions of the LEGO figure's neck and arms. You will also need a nice pair of fabric scissors to cut the fabric to shape.

**TIP**: Fabric scissors should only be used to cut fabric. Trimming other items with them

*Leather punch.*

can damage them, thus when cutting cloth with these damaged scissors additional fraying may occur.

Cloth is cheap and can be found at any fabric store. Remember you don't need much for minifigures; so check the scrap and discount areas. Another alternative are pre-cut quilting squares available at many stores. Just make sure you are buying broad cloth, if you want to stay with a similar fabric to what LEGO uses. You will also need an anti-fraying solution, which is also available at the cloth store.

Alternatively there is a product called printable fabric. There are a wide array of manufacturers and great variability amongst them. This cloth can be used to print designs on your capes and other cloth items. An inkjet printer is used to print on this material. This material is commonly referred to as printable fabric or photofabric. The biggest limitation here is that the back side is typically white after the backing paper is removed. It is the backing paper that allows the cloth to go through the printer. There are methods to make your own printable fabric, which will be covered elsewhere (page 36).

| Cloth Accessories | | |
|---|---|---|
| **General** | | |
| Leather Punch | www.harborfreight.com/leather-punch-tool-97715.html | $7 |
| Scissors | | $5 |
| Printable fabric | http://www.amazon.com/s/ref=nb_sb_noss_2?url=search-alias%3Daps&field-keywords=printable+cloth&rh=i%3Aaps%2Ck%3Aprintable+cloth | |
| **Total Cost** | | **~$12** |

## Photography/Digital Effects
## General Tools (Light Tent, Backgrounds, Digital Editing)

Please visit the Digital Photography section (page 68) to find out how to make your own economical light tent. If building one isn't your cup of tea there are many versions available online for a wide range of prices. Many include lights and backgrounds, so make sure to look for a deal. As for digital editing Adobe Photoshop is the standard. However, it is hard to beat the free Irfanview program if you are on a tight budget as it can be used to do many editing functions.

So if you want to dive completely into the hobby with the best options as I see them you will have about $220 worth of items to purchase including a printer. I don't recommend starting the hobby by purchasing every item; figure out what you really want to try first. Work on the skill set required for that area of the hobby and once you have perfected it, then move on to another area. I would buy the least expensive items first and then move on to more expensive options as you improve your skills. You don't need expensive equipment to get great results, mainly just time and practice.

| Photography/Digital Effects | | |
|---|---|---|
| **General** | | |
| Ifranview | www.ifranview.com | Free |
| Light Tent | Photography section for DIY | |
| **Total Cost** | | **Free** |

Designing art for minifigures sounds easy: draw a picture of what you want your custom figure to look like using a computer. While this sounds easy, there are many aspects that are critical to achieving quality results. Most of these aspects stem from the limitations of applying or printing the art onto the figure. The aim of this chapter is to get you to think about the end result while creating the art. To begin this topic, we need to understand how different art programs create graphics, specifically raster and vector formats.

## The Foundation: Raster vs. Vector

Raster image formats are made of tiny squares of color called pixels; these are primarily used in digital photography. The main graphic formats of the web, GIF (Graphics Interchange Format) and JPEG (Joint Photographic Experts Group), are raster formats. When you zoom in, these images become blocky (or pixilated). Think of a LEGO mosaic when you can see a raster picture, stand far away you see a picture, stand close you see squares. This image format will always have the limitations of the pixel size and pixel color.

Vector formats are not based on a square pixel but mathematics, as such, images will never appear pixilated. EPS (Encapsulated PostScript), to some extent PNG (Portable Network Graphics) and native formats like AI (Adobe Illustrator) and CDR (Corel Draw) are vector formats. When you magnify vector art, it stays sharp and clean because the same math applies at whatever the magnification. Think of looking down railroad tracks, they never meet and will never meet, even using binoculars, you maintain resolution despite the magnification. In a raster format, because of the pixel size limit the tracks will meet when magnified. Vector graphics are used in illustration and design (commercial artwork), so most home users doing simple web graphics, drawing pictures, or photo editing don't have a need for these formats. However, for the best results it is important that you create your designs in a vector art program. If you don't have a vector program, consider a trial version of the aforementioned programs or check out DrawPlus 4 or Inkscape (check the Toolbox chapter for links, page 14), which are free. If you don't want to use a vector art program, raster programs like Adobe Photoshop can be used, just remember to set your dpi (dots per inch) as high as possible. This value is the resolution of your image, thus the size of your pixel, which is tied to its print quality.

*Raster Image: All digital photographs are raster based images, if you zoom enough or the dpi is low enough the pixilation will be visible.*

## Color

Color is a complex topic and will affect how your art appears on the computer screen versus how it appears when printed. To understand the differences between the monitor and the printer let's consider the two formats used to create and display color, RGB and CMYK. These two formats result from the difference in the behavior of light mixtures (additive color) and pigment mixtures (subtractive color). What this means is that color is created by two very different phenomena. When light is perceived by the eye we are examining the wavelength of the light in the visible spectrum that is reflected by the object. When this light is separated by a prism you can see all the colors that add (additive color) together to make white light. As you move

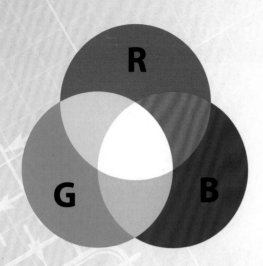

RGB (Red, Green, and Blue) values are mixed by addition to create all other colors. These are primarily used in TVs and other display devices where they represent small color elements that when mixed make every other color. Black is achieved by the absence of all three colors, where as white is achieved by mixing red, green, and blue in equal value. RGB color format is most commonly used in raster imaging and will look best when presented on a display device like a monitor, projector, or television.

CMYK (Cyan, Magenta, Yellow, and Black) values are mixed by subtraction to create all other colors. These are used in printing devices and benefit from adding a true black color. CMYK color format is most commonly used in vector imaging and will look best when printed. Printers predominately use the CMYK format, thus if you create your graphics in a CMYK format the colors don't have to be converted from one format to the other before printing. This format doesn't always look the best on the screen, as the colors are converted before displaying in RGB, but the goal is the final printed version, not the intermediate.

up the wavelength you proceed through the colors (Violet, Indigo, Blue, Green Yellow, Orange, and Red). When you see color from an object, it is the result of the reflection of light off that object. Specifically, all colors except the reflected color are absorbed by the object, thus this is subtractive color (all colors are subtracted from the resulting color as they are absorbed by the object). Therefore we have two formats to use create color, RGB and CMYK. To understand why this is important consider what how each are used.

Thus when choosing a color format, weigh the objective: in this case, it is printing your minifigure designs. Check your printer, most use CMYK. If this is the case, design in this format. Luckily the CMYK and RGB color values are published on Peeron for the LEGO color palette (www.peeron.com/cgi-bin/invcgis/colorguide.cgi and www.peeron.com/inv/colors). LEGO even released a small version of their palette (http://www.brothers-brick.com/2010/01/28/LEGO-releases-its-internal-color-palette-news/). If you utilize colors from the LEGO color palette, any design you create will appear at home in your LEGO structures and vignettes. Remember that every printer is slightly different and any color value used from the published palette might not be exact, but it will be very close. As such, small test printings to perfect the color match can be performed. Remember to check the color once it is applied to the figure, as application could alter the color slightly.

## Minifigure Art Design Guide (LEGO Style)

Based on an intensive study of official LEGO figures I have compiled a set of guides for designing art for the LEGOverse. This guide is in keeping with the LEGO style; I offer it as merely a guide. Please develop your style. These rules are not absolutes and I find that I go against this guide regularly in my own work. These are simply observations I have made based on examining what LEGO does and doesn't do. Most often what LEGO does and doesn't do is based on the print technology they are using to achieve the best looking figure possible. It is this detail that is the most important about this guide.

LEGO designs have been growing in complexity but are still limited by the technology LEGO uses to apply the art to the figure; typically LEGO uses pad printing. This is a great process to apply ink to a complex surface such as a helmet, but it is also expensive for one-off figures and thus limited. Pad printing is performed using an indirect offset printing technique where an engraved silicon pad is inked and the ink is then transferred via a 2d pad to the 3d object, one color per pad and multiple pads per design. The pad stretches and contours to the 3d surface. The limitation here is the creation of the pad is specific to a single design and a different plate and transfer pad is required for each color. This makes this method practically impossible for the customizer: it doesn't lend itself to a wide variety of figure designs. However, it does explain where LEGO applies details and where it does not. Typically LEGO applies art to the flattest portions of the figure. Have you ever seen a LEGO arm where the art was wrapped around the arm? No, it simply sits on the outer, flatter, region. The lower portions of the legs are also not printed because the foot is in the way. Therefore, LEGO typically alters the element color for the foot and prints the upper portion of the leg. Keeping this in mind when you design your art is critical. You have to understand the technique you are going to use to know its limitations so that these limitations don't ruin or control your vision.

LEGO also limits the number of colors in any one design to 5 or less. This is for simplicity and cost. Think about it and go look at your official figures. Bet you can't find one with more than 5 printed colors. If more colors were used the registration (alignment of all the colors against each other and the figure) of all these colors gets complex and more difficult, resulting in more print failures. The tolerance in a figure requires extreme control and accuracy, or a trick to hide errors (more on this shortly). This is a small space and in the vector program it is easy to clutter because it is so scalable. Keep your art simple and choose your colors well.

Let's talk about that trick while you have your collection of official LEGO figures out. Look at how LEGO details figures, specifically how LEGO separates one color from another. LEGO is being tricky and smart at the same time. Typically any two colors in any design are separated by a dark colored line, typically black. This line is used to visibly separate the colors and it also defines the smallest details because of the line weight or thickness (stroke, think of a line drawn by a #2 pencil compared to a mechanical pencil). I believe the line thickness used is approximately 0.6-0.7 stroke thick (this value will be in your vector program of choice). You can only make a circle so small when using this fixed line size, thus this limits the detail level of your art. I have noticed that thinner lines are used in LEGO design when they appear against a single color; I assume this is because of the lack of critical registration due to the single color. Now that I have explained how LEGO uses line weight, let's think about why. This will be important for your work as it is to LEGO's. When LEGO prints the art on figures it has to register the colors against one another (recall from above, each color is separately printed in sequence using a transfer pad). In order to do this there are certain tolerances, which are hidden beneath the dark separation lines, so they are covering their error. Even though an inkjet is more accurate, it is still based on a liquid droplet and will have bleed. This bleed can be masked under a separation line, so I suggest following LEGO's example.

As with any guideline there is an exception. LEGO typically outlines the edges of design elements (clothes, belts, etc.); however there are limitations to these outlines. LEGO only outlines these details where they meet other details, but not at the edge of the figure, thus belts are not outline on the edge where it is wrapping around the figure, creating the illusion it is wrapping around the figure.

Recall from the chapter 2, *What is Customizing?* that you need to have content when you design a figure. Who is the figure you are creating, how much life can you breathe into the design? In order to do this LEGO also tries to avoid symmetry in its minifigure designs. Make sure the clothes you are designing for your figure to wear make sense for that figure. Is the character neat and clean or a messy troll? Would a troll have a big shirt stain or a hard-pressed collar? Give your characters some character, but remember that the figure body is small. Don't overly clutter your design, keep it simple.

In an attempt to make the figure appear more 3-dimensional LEGO has added shadows to many areas of the minifigure design. These can be difficult to see due to the outlining LEGO uses, but they are ever present. See the article on the History of the Minifigure in book one for more details. Just know adding small dark-colored shadows will give your design more depth. Another 3D affect commonly used in art is the use of gradients. Gradients

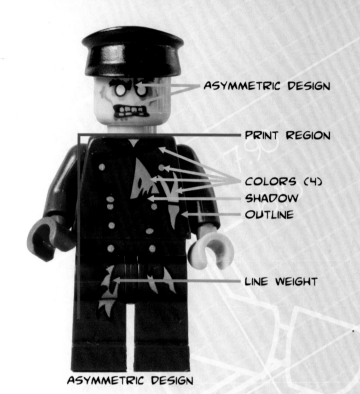

ASYMMETRIC DESIGN

PRINT REGION

COLORS (4)
SHADOW
OUTLINE

LINE WEIGHT

ASYMMETRIC DESIGN

*LEGO Design style.*

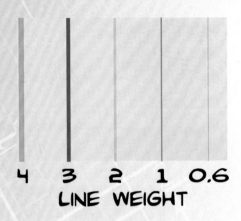

## LINE WEIGHT
4  3  2  1  0.6

*The above is a simple representation of line weights. Thinner lines are used for fine details, thicker lines for bold details.*

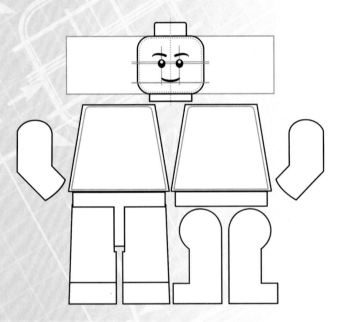

*The minifigure head is a unique body part that is out of scale to the rest of the figure. As such, designing appropriately for the head is quite difficult in relation to the rest of the figure. To begin, let's think about the scale of the minifigure face. The rule of thirds, a concept taught when sketching the human face still applies; basically divide the face into 3 equal portions vertically. This allows for the scale and placement of the facial features in scale to the head (Top of brows, Center of eyes, and Bottom of mouth). Designs featuring hair require the decal to flow onto the curve of the minifigure head to allow the decal to be properly placed in relation to minifigure head accessories, which would occupy the space above the brow line. Please see my template above. This template helps me keep the face in proportion to the LEGO head. I have placed several reference lines on the template; from top to bottom, Center of Eye Twinkle, Center of Eye, and Bottom of Mouth.*

are regions where a design gets lighter or darker through the use of small tiny dots called half-toning. More dots make an area darker, and fewer make it lighter. Gradients are difficult to use in minifigure design due to the print resolution and the ability to under-print lighter colors. Use gradients sparingly.

Two last words on minifigure design in a LEGO style, keep the art more cartoonish as LEGO avoids too much realism as it is a child's toy. Keeping this whimsical design feature will keep your figures at home in the LEGOverse. Don't make your designs too static. The figures you are creating are supposed to run around, fall down, and be pretty much like you and I. Give them small imperfections, wrinkles, and if you follow no other suggestion above follow this one, avoid perfectly straight lines.

## Remember Your Printer

As stated above, most of what LEGO does in the design process is specifically geared towards the method it prints figures. Please learn from LEGO's example and remember your printer/ print method. Most printers have difficulty in printing a line with a weight less than 0.2 stroke. Just because you can see the difference in your vector program doesn't mean you can print it, test printing is important. Vector programs can create art beyond the abilities of home (non-commercial) printers due to the nature of vector art. A good point size range for details is 0.5-4.0 pts.

## Templates

Now you understand image formats, color, and simple LEGOverse design the fun can really start. The first thing you will need is a template of the LEGO minifigure. With a template you will know the exact size to design any decal. To create a template, grab a metric ruler and a pen and paper. Merely take the part you are designing for and accurately measure the outside edges. Then redraw this structure in your art program to the exact dimensions. While this sounds easy it can be challenging due to the curvature of certain body parts. These surfaces can be measured using fabric tapes. Another trick is to place the body part on a piece of paper and trace the part; measure the resulting tracing.

By creating everything to scale it avoids conversion errors when it is time to print. The basic template is at the left. Please note that the decal area of the torso is ~1.42 cm x 1.2 cm, the hips are 1.4 cm x 1.1 cm, and the legs are 1.42 cm x 1.2 cm). Many other templates can be downloaded at a site I helped found, Minifigure Customization Network or MCN (*www. MinifigCustoms.com*).

## Advanced Theme Design

This section will examine how characters are designed to fit into a cohesive theme, specifically how to personalize characters into a group and not create repetitive figures. With the information presented, one could construct an army, pack, clique, unit, or whatever, personalizing each to make them unique, yet allowing for some similarities so they fit into that unit, army, world, or universe. This section has been inspired by a great comic created by Michael Anthony Steele and Scott McFadden who are experts in creating a theme, which is why I chose this comic as subject matter for this section. Anthony and Scott, of Big Red Boot Entertainment, are the creators of the digital comic book called *Clown Commandos*. The examples in this section merge the *Clown Commandos* main characters with the LEGOverse (so technically the figures presented are designed in two themes). To do this well one has to stay true to both the comic and to LEGO style, this way the characters are recognizable and yet don't readily stick out in either world (comic or LEGOverse).

The premise of *Clown Commandos* (CC) begins in an alternate universe where a single occurrence during World War II launched a chain of events that caused clowns to evolve into an elite fighting force. In present day, a squad of Clown Commandos gets trapped in our reality, where the rubber nose is NOT a badge of honor.

Part G.I. Joe, Ninja Turtles, and Batman, these juxtaposed heroes fight crime while searching for a way home. To them there's nothing funny about being a clown, the heart of the series' humor. No matter what kind of life or death struggle, no matter how evil the villain, Clown Commandos think nothing of taking down the bad guys with a well-aimed rubber chicken or a pie to the face.

Before merging of the *Clown Commando* world with the LEGO to create the figures in LEGO form we need to understand how the Clown Commandos were artistically created. By examining themes in this way a better understanding of unit/group character design can be developed and used to create more interesting/cohesive groups. This gives the figures more life, more back-story, more reasons to have that blue hair, seltzer water pistol, or red floppy boots. For Scott's (artistic lead) insights please see the inset on the next page.

When designing in a theme, there are critical elements that need to be repeated to make the characters similar such that the theme is supported, yet the characters need to be different enough to make them individuals. These critical elements can be reused from figure to figure so long as some twist on the element is utilized to make it new and individualize it on each character. Without individuals the depth of the unit is lacking; each figure needs to have that personalized stamp or they merely look mass produced or robotic. The point of customizing your world is to make that unique character. If sheer numbers is your game, exact copies are fine: think stormtroopers when they line up in the movie shots in phalanxes (100 soldiers in 10x10 rows and columns). However, if you are creating that small fighting unit or even that garage full of mechanics having them all in the exact same outfits with oil stains in the same place gets tedious. Think about the Ghostbuster's films, did all 4 main characters wear the exact same jumpsuit the same way? No, of course they didn't, they all added their "personalized flair." Maybe not 27 pieces, but I digress. Think about adding "flair" to each character, make them unique, and recycle design elements in unique ways.

Cpl. Honky: CC Quartermaster

Pvt. Jumbo: Engineer Clown

The Clown Commandoes.

Sgt. Blammo: Explosives expert of the CC.

*Clown Commando Humvee: Adapted design from Dan Siskind of Brickmania.*

## Scott McFadden (Artistic Lead) Speaks!

Clown Commandos, now there is a dichotomy. Opposites are supposed to attract, right? Well that is what we had to figure out. We wanted to find a way to combine the humor and lunacy of a clown with an elite fighting force. Like our concept, we wanted find a way to take these guys seriously. So we started researching both sides. We started with a more military/mercenary look (beyond that of the enlisted ranks) and began adding clown elements. And, like our concept, we explored how these elements could evolve through the years and merge (and actually make sense, oddly enough). The polka-dotted camo is a good example of that. Of course this is how clown camouflage would evolve!

When it comes to the Clown Commandos' alternate universe, we have strong art concepts, too. Since the clowns of their world no longer entertain, much of the laughter and joy has gone. We symbolize this by the darker palette of their world, as seen through flashbacks. We keep the palette more desaturated and gritty, giving the feel of a police state/Gotham City look. Thus, once our heroes pop into OUR universe we have a brighter palette and a friendlier world.

Each character has its own personality and we tried to capture that with their look and uniform. For example, Lt. Bubbles is more of a rebel yet she still has the pride of the CC. We tried to make each character unique and their place in the squad somewhat apparent by their appearance.

On weapon design, we took advantage of the fact that this is, after all, a science-fiction concept so we utilized "clown tech" or "mime tech" to merge the funny with the military. We try to keep them a little intimating, but with touches of primary colors so the joke is subtle. Plus, the non-lethal factor plays through as the weapon is fired: a pie shooting out of a pie gun, an over-sized bubble gun that explodes with sleeping gas, and of course, there's Colonel Clown's regulation seltzer pistol. Again, military first, the pistol is a normal military-style pistol with a seltzer chamber attached.

When we started concepting for this project we wanted it to be action-adventure premise that just happens to be a comedy. We built everything with a strong military foundation, from the weapons to the uniforms and even the ranks. For instance our colonel rank replaces the real rank's eagle with a rubber chicken. With almost everything, we soften the edges a bit and added a primary palette. That is how we try to handle all our elements, military first with a touch of funny.

The key to creating the squad in LEGO form is their badge of honor, the red nose. I utilized the nose to make these figures unique and yet totally LEGOesque. The nose was constructed by cutting the tip off of a LEGO antenna element. The LEGO head had a small hole drilled in it to allow the antenna shaft to fit inside and hold the nose onto the minifigure head. The unit was also further identified as clowns by my sculpting the big red shoe elements making these clowns feet twice as large as a normal minifigures. These two custom elements were continually utilized as baseline for all clown characters, except one (please note as in life, there is always one exception to every rule).

Colonel Clown is the leader of the Clown Commando unit. Instead of donning the whole uniform he fights in a tank top, yet still wears the Clown-camo pants, as seen below. His prominent dog tags describe this clown as military issue. His arm bandolier and wrist guard further set him apart from his squad mates (watch that bandolier: this would be one of those elements that is recycled in new ways). He is geared up for non-lethal combat with his happy bomb (exploding in rapidly-expanding knock-out gas), seltzer pistol, and seltzer cigar. The happy bomb is a painted and decaled BrickArms product and the seltzer pistol is a modified BrickArms 1911 that has the addition of a piece of trans-blue LEGO bar. The seltzer cigar was created by using the LEGO antenna element shaft; specifically painting, cutting to length, and gluing (cyanoacrilate) to the minifigure head. The LEGO hairpiece was custom painted bright green. To further make Colonel Clown more LEGOesque I designed his tank top and musculature after an official LEGO mechanic minifigure, this helps blend the comic style and the LEGO style together. This tank top design with a few tweaks was also used to create Lt. Bubbles, seen at the bottom.

*Colonel Clown in minifigure form, with his comic form on the left and minifigure torso above.*

*Lt. Bubbles.*

When creating the rest of the squad and other characters from the *Clown Commandos* comic, several design elements were merged from the LEGOverse with that of the Clown Commandos. For example, the Ringmaster's and Mr. Smiley's suits are modifications of an official LEGO suit design. I also used the standard LEGO Mime for the Mime Syndicate members to help blend the worlds together. Another trick used was whenever possible I used official LEGO elements and modified them for their specific purpose. There are several examples of this: Jumbo's helmet is from the Toy Story Army Men, Honky's pants are a modified LEGO element, and Blammo's hat is a modified LEGO crown. I also used a LEGO solution to achieve Jumbo's enhanced size by using the LEGO hockey armor element and the BrickForge ammo pockets. I also boosted his helmet to give him just a bit more height.

Creating or merging two worlds can be done in very imaginative ways. When merging the Clown Commandos with the LEGOverse I tried to stay true to both creating new elements only when essential and using already accepted aftermarket elements (BrickArms weapons and BrickForge accessories). Where does your interest take you? How can you merge that interest with the LEGOverse?

*Clown Commandos images courtesy of Big Red Boot Entertainment. © 2011 Big Red Boot Entertainment LLC, all rights reserved.*

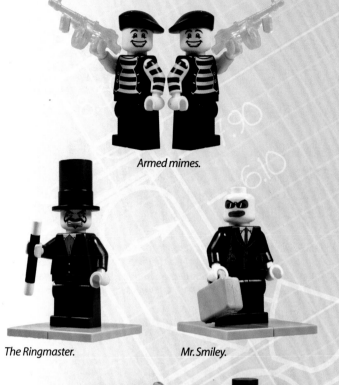

Armed mimes.

The Ringmaster.                    Mr. Smiley.

## How to Get Art on a LEGO Minifigure: Decals vs Direct Printing

Now that we have had an extensive discussion on designing, design styles, and designing in a theme you should have created art to construct several custom figures. How do we get this art from the computer and onto the minifigure? There are several methods waterslide decal film, sticker film, static-cling sticker, pad printing (previously discussed), or inkjet printing. Most of these techniques have some major limitation. So you have to ask yourself, what are the goals for the figure you are creating? Is it a collectible requiring the highest quality or is it something for your 5-year-old cousin to play with? Waterslide decals can be fragile if they are not properly sealed and even then can be damaged in hard play, yet they give the best appearance when applied to the figure and are inexpensive and easy to create. Sticker film and static-cling stickers will give a thick appearance and it will be very apparent that something is stuck onto the LEGO torso. However since these are more hard-wearing and much easier to apply, maybe this is the choice for that cousin. Plus, in the case of the static cling, they are easily and rapidly changed (look to Brickstix: *www.brickstix.com*). Pad printing was discussed above: while it is expensive and limited in the ease of creating a wide range of designs, but it is likely the hardest wearing, which is why LEGO uses it. The final is inkjet printing. This requires a special direct to substrate flatbed printer. These are still quite expensive ranging from several thousands of dollars to more than $30,000. This is another versatile and hard-wearing option, but not as scalable as pad printing, which is why LEGO doesn't use it for production sets. What is right for you? For the price and quality of the finished figure, I recommend waterslide decals.

The Mime Syndicate.

Col. Clown on the road.

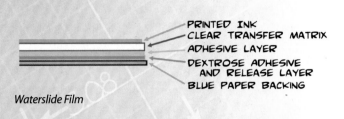

PRINTED INK
CLEAR TRANSFER MATRIX
ADHESIVE LAYER
DEXTROSE ADHESIVE
AND RELEASE LAYER
BLUE PAPER BACKING

*Waterslide Film*

*ARF Digital Camo Clone Troopers. Helmets by Arealight. Note the red and tan figures feature pad printed helmets, all the others were created through the use of waterslide decals. I designed and created the decals, which were then applied by Michael "Xero" Marzilli.*

*Legend of Korra minifigures with decal details.*

# Waterslide Decal Film: Technical Info

With the advent of printable film for inkjets and laser printers, as well as color copiers, custom decal creation at home using waterslide film has become accessible. Waterslide decal film, or water slip, is a thin film used to transfer a design onto a surface using water as a transferring agent. The design is printed face up on the waterslide film, which relies on a dextrose (from corn) to release the film from the decal paper. The dextrose also assists in the bonding of the film to a new surface. Quality film also features a water-based adhesive layer to improve the bonding of the film to the surface. The film itself is typically made of lacquer.

When considering the various types of materials one can print on and use in the creation of custom figures waterslide film has a massive advantage over all the rest: its thickness. Because waterslide film is much thinner than the others it is capable of taking on complex curves, something that many other decorative techniques such as vinyl stickers and static cling stickers are not. This is also a challenge to some direct printing techniques (discussed earlier in design section). As waterslide film is printed, it can be produced with very high levels of detail. Because of all of these characteristics, waterslide film is the method of choice when it comes to customizing figures.

# Decals and Printing

This section will begin with the best advice I can offer: *before printing on any expensive media always print a test page.* Print your designs out on a piece of scrap material or paper. Confirm your color choices as colors on your screen will not exactly match printed colors (RGB vs CMYK typically, see color section page 16). Confirm your details, as you have been working on a magnified design in a vector art program that is actually quite small when printed. Are your details too fine to print or is your design overly cluttered? These questions can typically only be answered once you see the printed product.

We just discussed that waterslide decal film is not only very thin and clear, it (when applied correctly) appears as if it the design was printed directly on the figure. The result should resemble the appearance of a LEGO printed element. Waterslide film is available from hobby stores or online. I recommend the film sold by Micromark (*www.micromark.com*); they have a sample pack of clear and white film. Clear film will work well on any of the lighter colored LEGO elements; however dark elements create a problem which requires you to use a special printer or use white decal film. This is because the darker LEGO element will show through the printed regions darkening them and in many cases completely concealing your design. This is due to the fact that the inks don't have the opacity to stand out on the dark elements, because they were designed to print on white paper. If you use white film this gives your design the ability to keep the vivid colors on dark elements, however you will now have to print the torso color or closely trim your decal. Printing the torso color will require a close color match to the torso (or the decaled element), be sure use to use the references listed on Peeron and Bricklink mentioned in the color section (page 16).

As most people have access to inkjet and laser printers or color copies (Kinko's, etc.), make sure you select the waterslide film that is appropriate for your printer or copier. Print your designs using the highest resolution possible for your printer (just like you did when performing your test print). Once printed be very careful not to handle the sheet of decals until the ink dries. This may take some time (up to 24 hours), especially if using an inkjet. The decal film isn't absorbent; the ink sits on the surface. This means smudging is very possible, so be patient, overnight drying is best. After you have printed using an inkjet or color copier, you must add a clear overcoat to the decal film; a clear spray paint available at any home improvement store. I recommend a Krylon UV-resistant clear spray paint. This is required because the ink is NOT waterproof on decal film regardless of what the printer's manufacturer states about their inks. Apply several thin coats of clear sealant and allow it to dry between applications (2-3 applications normally does the trick). This will protect the ink from the water used in the application of the decal, so use of a sealant is critical! Once printed and sealed the decals can be cut from the page for application.

*Avenger minifigures with decal details.*

## Decals: Advanced Application

As the focus of this book is more advanced techniques, knowledge of basic application is assumed. If you need a review please see the first book, my website (www.fineclonier.com) or my Youtube account (http://www.youtube.com/user/FineClonier). With an understanding of the basic instructions you have the foundation for the advanced instructions. The basic is great for decaling torsos and heads; however it can get much more difficult when you attempt helmets, shoulders, and custom parts. This is because flat surfaces are pretty easy; complex curves get much more difficult and require solutions to help the film conform to the curved surface.

*ACU.*

*Shaak Ti.*

# Advanced Waterslide Decal Application Instructions

Completely disassemble the minifigure.

Remove any printing on the parts using Brasso that you wish to decal. Wash the figure parts to remove any residual abrasive.

Trim the decal with scissors.

Using the tweezers, dip the decal into the distilled water.

Allow the decal to sit for 60 seconds to allow the decal glue to release the backing paper.

During the 60 second interval apply Micro Set (setting solution) to the part to be decaled with a small brush to prepare it for the decal. Then slide the decal from the paper backing onto the Micro Set wet application surface. Position the decal into place with a wet cotton swab.

Gently roll a moist cotton swab over the surface of the decal to remove any trapped air bubbles and excess water or Micro Set. If the decal shifts slightly during this stage reposition and allow the decal and torso to sit for about 10 minutes and partially dry.

Apply multiple coats of decal softening solution (MicroSol) waiting approximately 3 hours between for evaluation of decal application. Once decal has set down onto the complex curve stop use of softening solution (4-5 coats max) and allow the decal to dry.

Overnight drying is required before sealing the decal (Check the tips below to speed this up).

Seal the decal with a clear spray paint, I recommend Model Masters Ultra-Clear Gloss or Semi-Gloss Lacquer.

Advanced decal application starts with using a chemical kit to improve the adhesion and contouring of decals to complex curves. These kits help make the decal crystal clear when applied properly, thus making the design appear painted/directly printed on the figure parts. Kits can be found at many hobby stores or online (Micromark, *www.micromark.com* or Microscale, *www. microscale.com*). If you are from a foreign country you will have to find these application kits in your country as they contain a mild acid that cannot be shipped internationally. I recommend Microscale application solutions: decal setting solution and decal softening solution. Decal setting solution (MicroSet) strengthens the bond between the decal and the part surface. It also helps remove any trapped air by preparing the surface for the decal. Softening solution (MicroSol) is for

*Brasso abrasive.*

*Model Masters
semi-gloss laquer*

*A work space with Microset, Microsol, cotton swabs, tweezers, water pan, decal application kit and brush.*

the most difficult irregular surfaces you find on LEGO parts (helmets, backpacks, etc). It completely softens the decal allowing it to drape down onto the surface of the model conforming perfectly and without distortion. Apply the solution generously and wait about 3-4 hours between applications. You will notice that the decal will wrinkle slightly. Leave it untouched as it is very fragile during this stage, it eventually will lay flat. If it does not lay flat, apply again. For very complex surfaces this may take 4-5 application spread across 12-15 hours. It is worth the time as the result will amaze you. The home-spun alternative to decal softening solution is white vinegar; use it as described above.

## Tips and Tricks for Application:

There are several tricks that can be used at every step of the basic and advanced decal application, so be sure to read the following and think about where to use them above.

### Cotton Swab:

The power of the cotton swab has been mentioned, but I really can't stress this tip enough. The cotton swab, especially the wooden stick version, is the best tool to apply decals. Use a very wet cotton swab when positioning the decal. However, a slightly damp swab is called for when removing trapped air bubbles and to absorb excess water or setting solution. A completely dry swab is never recommended as it can stick to the decal. These wooden tips of these swabs are also useful to position the decal (See the hair dryer trick section following for more cotton swab uses).

### Trimming/Strategic Cuts:

After you have read the advanced decal application and the use of decal softening solution you are likely asking, "Why do I need any type of cut strategy?" Well through the use of strategic cuts you can help the decal conform to curves and help remove the possibility of wrinkles in your decals. Wrinkles can occur when using the softening solution, especially if you get impatient and try and help the decal conform to the part surface. While decaling a round surface with a gentle curve make small slits in the edge of the decal once it is on the surface of the part. Do this with an X-acto knife or razor blade. Shoulders and shields are great examples of when this is helpful. This will allow the flat decal to curve and overlap slightly to take on the desired shape. This strategy can also be applied to the head tails in the advanced decal application example.

### Softening Solution Alternative:

As previously mentioned, white vinegar can be used in place of decal softening solution. To make this solution add 1 part vinegar to 3 parts water. Use this solution just as you would use decal softening solution.

If you are getting impatient for your decals to conform, you can also use a damp cotton swab to help the softening solution along, but honestly it is best to be patient. Whatever you do, DON'T use your finger as you will leave fingerprints in your softened decal which will ultimately detract from your finished figure.

### Hair Dryer:

Another trick to speed the process along is the use of a hair dryer, which can be used at every decal application step, including the basic steps to help speed evaporation. Just be careful to remove all trapped air bubbles before using it. This will also help the decal conform to a curved surface because of the heat. It is best to use it on low heat and low air speed, if possible.

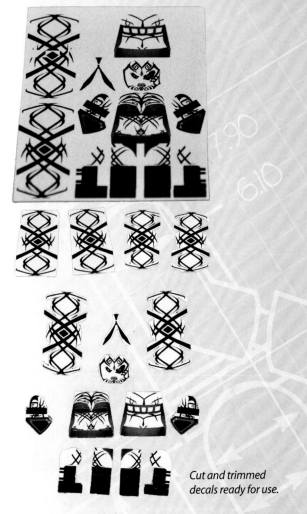

*Cut and trimmed decals ready for use.*

*Application of decal softening solution (To see the complete application of the decals used in Figure 1 check this Flickr set: www.flickr.com/photos/kaminoan/sets/72157605952528360) Custom head piece by Bluce "Arealight" Hsu. – Photos by Jared and Amber Burks.*

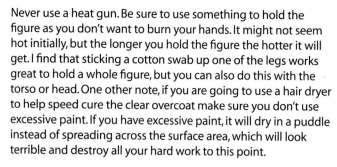

Never use a heat gun. Be sure to use something to hold the figure as you don't want to burn your hands. It might not seem hot initially, but the longer you hold the figure the hotter it will get. I find that sticking a cotton swab up one of the legs works great to hold a whole figure, but you can also do this with the torso or head. One other note, if you are going to use a hair dryer to help speed cure the clear overcoat make sure you don't use excessive paint. If you have excessive paint, it will dry in a puddle instead of spreading across the surface area, which will look terrible and destroy all your hard work to this point.

## Leg Application:

When applying a decal to the minifigure leg, decal damage can occur if the leg is bent into the seated position after application. This is due to the torque applied to the leg, which presses the leg into the underside of the hips. As there is a ridge in this area, it scrapes the decal off the leg's surface. If this ridge is removed or at least diminished, the odds of this type of damage occurring is reduced. This ridge can be removed by scraping it with a hobby knife from the inside of the leg area to the outside. Also if one carefully places the leg in the up position then the leg is not pressed into the hip and damage is less likely. Use of the advanced application also helps as the decal more tightly bonds to the leg's surface, which means it will rest lower and make it hard to scratch off.

*Application of decal softening solution and the final figure is shown below. (To see the complete application of the decals used in Figure 1 check this Flickr set: www.flickr.com/photos/kaminoan/sets/72157605952528360) Custom headpiece by Bluce "Arealight" Hsu. – Photos by Jared and Amber Burks.*

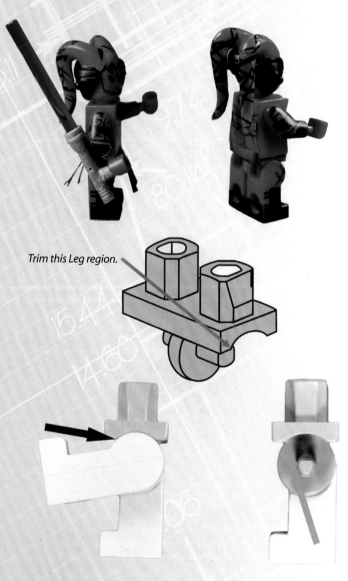

Trim this Leg region.

# Rendering and Virtual Customization

Sometimes the best looking designs just "don't look right" when you finally get them on the brick. Sometimes it is the design, sometimes it is element, and sometimes it is the color of the brick. One way to get a better feel for the look of the final product is to render it beforehand. A common, relatively easy, and free rendering option is to use LEGO Digital Designer (LDD) and POV-Ray with the LDD2POVRAY conversion application. LDD2POVRAY (*http://ldd2povray.lddtools.com/index.php*) converts any LDD model files into the POV-Ray file format. This allows for the generation of photorealistic LEGO designs.

Rendering with LDD2POVRAY and POV-Ray gives a much better feel for the actual brick colors and lighting. Rendering does take time however. A modern computer could take an hour or more to generate a single figure. Thankfully, POV-Ray allows you to queue up several renders to process sequentially overnight.

Translucent elements should be avoided though as even a single small element can easily triple the processing time. This is due to the complex math required to apply refraction on the elements "behind" the translucent part.

Once your render is complete you can use image editing software, like Photoshop, to virtually apply your prospective decal designs.

Virtual customization refers to the art of applying your minifigure art over a LEGO figure in a photo using a program like Photoshop. You are creating a virtual custom figure. Many figures have been created using this method and several artists are using animation style figures to make their figures look more like the way LEGO's print and media advertisements. This is a developing area of the hobby. While the details are beyond the scope of the book it is relatively easy to add the art over figure parts. The cover of the first book and elements of the cover of this book were created with this technique.

Here is an example of a variety of figure templates Glen Wadleigh tried out for his custom Codex figure from *The Guild*.

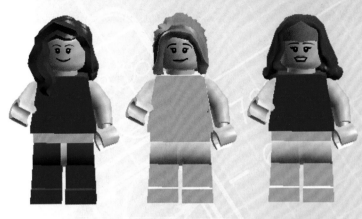

*Rendering steps to create Codex from the Guild by Glen Wadleigh. LEGO Digital Designer (LDD) is easy to use and even without hi-quality rendering can make effective templates for basing your designs on.*

*High Quality Renderings of the figure bases for Codex by Glen Wadleigh.*

*The characters from the* Guild *by Glen Wadleigh.*

Custom minifigure of Carly Foulkes. The black helmet began life as a pink one. Using a black sharpie the helmet color was altered around the center stripe to make it black.

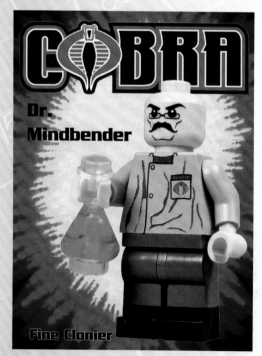

Dr. Mindbender's boots were added using a black sharpie. This is faster and easier than applying decals to these regions; it is also more hard wearing.

Decals, stickers, or direct printing isn't the only way to add details or color to your custom figure. We can revert to the old-school painting techniques that have been around for miniature decorating for years. I will give you one word of advice, *practice*. Painting details by hand requires a patient touch and a steady hand. Small brushes and toothpicks help too.

Back when LEGO was in serious financial trouble, the company instituted a part cap limiting the number of elements that exist regardless of color. This was done specifically for financial reasons. They have to create, store, manage, and fill boxes with all those parts. The fewer the parts, the easier this is to manage. LEGO maintains part warehouses with robotic storage systems with 500,000 storage slots. While this is exciting, LEGO doesn't maintain 500,000 different parts in constant production. With every new part created, an old part must be retired. There are about 2,350 different elements in the LEGO range – plus 52 different LEGO colors. Each element may be sold in a wide variety of different colors and decorations, bringing the total number of active combinations to more than 7,000. While this seems like a large variety, most are bricks and not for minifigures or minifigure accessories. This means LEGO can't make everything we customizers need or want. This is why the grey market for LEGO accessories exists and why it is growing, but I digress. For two great articles on the topic of what and why LEGO has these limitations please see: *http://bricks.stackexchange.com/questions/1889/why-does-LEGO-restrict-the-number-of-colors-per-shape* and *http://natgeotv.com.au/tv/megafactories/*

It has just been explained that LEGO can't make every part in every color you desire, so let's explore how to get that part in your desired color. There are some easy ways to alter the element's color or add details to official parts. Briefly, this chapter will go through the different options for altering a part color; markers (Sharpie®), paints, fabric dyes, and vinyl dyes. Altering part colors can be messy, so please use care not to alter the color of your clothes, furniture, or most importantly, you. *Several methods including, but not limited to aerosol paints, fabric dyes, and vinyl dyes will require adult supervision, so please be sure to seek help if you are under the legal age to purchase these products.*

There are limitations to these techniques that will be described, so if you are concerned over the final result or have never tried a specific technique, please use a practice part. Be smart about your practice part: some elements are very expensive and hard to find. If you can't use a duplicate part, use a similarly-colored one to work out your conditions as the original part color can influence the final result.

## Part Prep:

To begin altering any part color, the piece must be properly prepared to get the best color adjustment. Preparing the parts is required to remove residues on the elements from their production. To remove the residues, scrub each part with a mild dish soap and water. Using an old toothbrush to get into the small cracks and crevices will help removal of all the residues. The next step is to dry the part with a soft towel or washcloth. To ensure the residues and water are completely removed, especially in the small crevices, wipe down the part with an alcohol wipe or

70% isopropanol (isopropyl alcohol/rubbing alcohol). The alcohol will evaporate quickly and help remove any water trapped in the crevices; it will also remove many non-water-soluble residues. Now the part is ready for whatever alteration method you choose to use.

## Sharpies

Markers (paint or ink - Sharpie) are especially good for small color changes. Broad coverage is a challenge and they are not permanent regardless of what the manufacture states. These are best for quick coverage of small areas that will not see much wear unless you clearcoat them. However, they can be used to alter part color for smaller parts and are likely the cheapest way to do so. If you use this method be sure to use even strokes across the part, and once dry, clearcoat the part with acrylic paint to protect the ink. If you are after a temporary color alteration, this is your method of choice. Sharpie ink can easily be removed with a touch of rubbing alcohol, unless the alcohol-resistant version is used (which has a special red label). The best use of a sharpie marker I have seen is to alter a light flesh minifigure head to yellow. I also commonly use Sharpies to add boot details to the bottom of LEGO legs. These are easily touched up in the case of wear and you can easily decal over sharpies once it is dry and clearcoated.

## Paint

Paint is the most widely used method to alter LEGO part colors. There are two types of paint: enamel and acrylic. Enamel is an oil-based paint that will dry slowly and requires paint thinner to clean up. Enamel can also have strong odors. If you are using it in any volume, be sure to take frequent fresh air breaks. Acrylic paints are water-soluble, meaning they are thinned and clean up with water, as long as they haven't completely dried. It is because of these two factors that I recommend acrylics.

I can not emphasize enough the power of primer. To achieve the best color result, primers are critical. I prefer Citadel's Skull White primer. Be sure to use light spray as the primer layer will be part of the paint buildup. When using paint, it is best to apply it in thin layers and build to the final finish. This will make a stronger finish overall and leave the least amount of buildup on the part. Buildup is the accumulation of paint on the part and when visible, it noticeably detracts from the final custom figure. It is better to apply three thin coats than one or two thick coats to an element to help avoid this.

There are two basic ways to apply paint: with a bristle brush or an airbrush (or spray can). If you have an airbrush (or you can find the desired color in a rattle aerosol can), you likely already know it is a wonderful way to apply paint to broad areas. Cheap airbrushes are all you really need like ones that use small compressed cans of air or a Preval Sprayer. Slowly sweep across the part applying three sequential thin coats of paint (allowing the paint to dry between coats). If however, you don't have an airbrush, adequate results can be achieved with a bristled paint brush with a bit of patience and practice. I prefer nylon-bristled brushes. Follow the recommendation above and use several thin coats. Be sure to give each coat plenty of drying time before application of the next coat. Painting takes practice, so you might want to try a few test pieces before painting your rare LEGO element. Having your paint properly thinned is also critical; this is easy when using acrylic paints. Simply remove a small amount of paint and place it in a small cup, add water slowly till it is easily and smoothly applied to the surface of the part you are attempting to paint.

*Orc Space Marine Captain by Michael Marzilli.*

*Hand-painted details add to Ghostbuster's figure, by Michael Marzilli*

If using spray paint I prefer one of the typical modeler's brands, Testors or Model Masters, but I have also found a brand called Montana Hardcore spray paint. This paint was created for graffiti artists and comes in an insane array of bright colors. Hardcore sprays better than any spray can I have ever used, which speaks to its use in graffiti art. If you can find it, it is worth the money and you will likely be able to find the best color match with the wide range of color options available.

If you can't find the right spray paint, there is a device that will allow you to turn any paint into rattle can. The Preval spray system (*http://shop.preval.com/collections/preval-sprayer/products/preval-sprayer*) attaches a small can of aerosol propellant to a paint reservoir allowing most any paint to be applied by aerosol. These devices are inexpensive and easy to use.

Paint does have flaws. Much like markers, it can wear or scratched off. It can also be hard to get a perfect finish with a brush and, as mentioned before, you can get paint buildup. Minifig hands are very difficult to paint as it will almost always chip off if you routinely place anything in the figure's hand. Because of these issues, many have sought out alternatives.

## Dyes

Dyes are more permanent color changes since they penetrate the surface of the LEGO elements. They do not build up on the surface making it impossible to chip them off the part. However, they can be difficult to find and temperamental to use. So there is no perfect solution to part color alteration. The two different dyes commonly used are RIT fabric dye and vinyl dye.

The RIT fabric dye website states, "RIT can be used to dye many different types of materials including wood, paper, plastic, feathers, and even canvas shoes!" RIT fabric dye can be used to alter the color of lighter elements to darker shades; however it cannot be used to lighten darker elements. RIT dyes can be mixed to create custom colors, making it a good alternative to paint if you need a darker element. RIT has massively expanded the information on their website about the various options of dying different materials.

**Microwave Dyeing Technique:**
*http://www.ritdye.com/dyeing-techniques/microwave*
Temperature helps drive the dye into the material.

**Dyeing color Guide:**
*http://www.ritdye.com/colorit_color_formula_guide*
Mix and match RIT dyes to achieve most any color.

**What May Affect Your Dye Results:**
*http://www.ritdye.com/dyeing-techniques/using-colorit-color-formula-guide*
Things to keep in mind when attempting to dye various materials.

**Tips for Success:**
*http://www.ritdye.com/dyeing-techniques/tips-success*
More suggestions from RIT on how to get the best results.

*RIT fabric dye.*

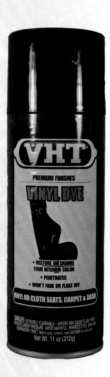

*Vinyl dye.*

## Another Perspective
## from Iain "Ochre Jelly" Heath
*50 Shades of Bley*
*http://tinyurl.com/50ShadesOfBley*

Typically it is best to experiment when using RIT dye. While we all think of LEGO as being ABS plastic, in reality LEGO uses many different types of plastics for their various parts. Therefore it is critical to test each type of part you intend to dye. If the part is ABS adding 20-40% acetone to the dye mixture helps the dye penetrate the ABS. Shortly the chapter will discuss vinyl dye, which also uses acetone to get a dye to penetrate plastic, low concentrations are safe for LEGO elements. At the 20-40% range the ABS should not be damaged, however, test each type of element you wish to dye. Again, LEGO uses several different types of plastic and they will all respond uniquely to the dye and the acetone. This is also aided by using water that is near boiling in temperature (140°F (60°C)). Just be careful, you don't want to melt your part (ABS plastic can be heated to 176°F (80°C), it will melt at 221°F (105°C)) and be sure to use a device (spoon) to help remove the parts from the dye. To begin, soak the part you wish to dye in water containing white vinegar for at least 30 minutes, but overnight is better. Again, I always recommend using an experimental part initially. Once the part has been soaked place it in warm dye (+20-40% acetone) containing white vinegar and check for the desired color every 15-30 seconds. Continuing to heat the dye helps, this can be done in the microwave or with a small plate warmer. It is easy to go beyond the desired color; frequent checking is best. Note the duration required to get the desired color and then dye your desired part for that duration. If you need to alter many parts to the same color, it is best to do so in small batches. Make sure to note the time and use fresh dye batches for each part to ensure consistent color. With the availability of 3d printed parts, this is an excellent method of coloring those parts. Why pay extra for colored material from the printer when it can be achieved easily and inexpensively from fabric dye.

*NOTE: There is a difference between the liquids and powders dyes. The liquids are not as concentrated and will take a longer duration to dye. The powders can be made more concentrated and will take seconds compared to the minutes that the liquids require.*

Vinyl dye is a difficult product to find, typically it can be found in automotive stores. Avoid the products called vinyl color (or colorant), which is easily confused with vinyl dye. Two common brands are VHT and Dupli-color (I prefer VHT). An alternative to automotive stores is custom automotive paint shops. It is VERY expensive to purchase vinyl dye from these stores, but when you need the absolute correct color this is the only option. Typically this product comes in a spray can, so be sure to use several light misting applications instead of one thick and heavy spraying. Let each application dry completely before applying the next.

Vinyl dye contains strong organic compounds which help the dye penetrate the plastic (similar to adding the acetone to the RIT dye); vinyl color (The product mentioned above that you don't want) is a fancy name for paint. If in doubt on which you have found, read the product components. One of the first compounds listed in vinyl dye should be acetone. Yes, we all know that acetone will melt LEGO elements when used in high concentrations, but in low concentrations it allows the dye to penetrate the LEGO element and alter the color, resulting in a

*Gollum by Iain Heath*

*Clutch Power's features a vinyl dyed hairpiece as LEGO doesn't have that hair style in black.*

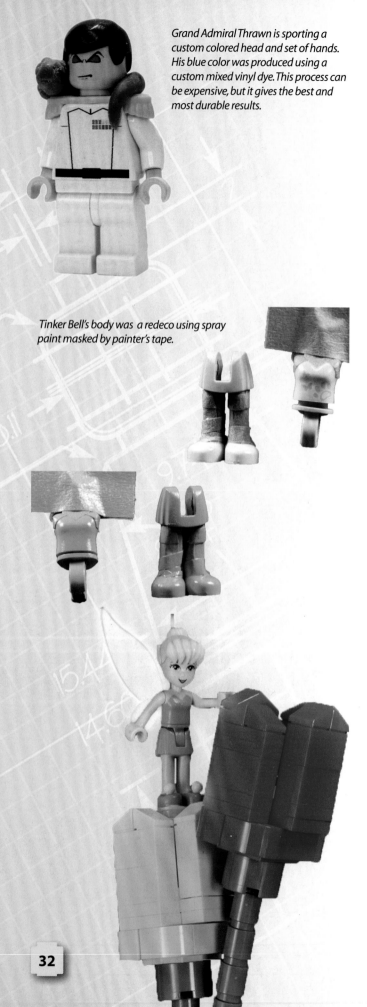

*Grand Admiral Thrawn is sporting a custom colored head and set of hands. His blue color was produced using a custom mixed vinyl dye. This process can be expensive, but it gives the best and most durable results.*

*Tinker Bell's body was a redeco using spray paint masked by painter's tape.*

permanent color alteration. There is no build up on the surface; true vinyl dye cannot be scratched off. Because of the mechanism of action, vinyl dye must be applied above 70 degrees Fahrenheit and below 85 degrees F. If it isn't applied under these conditions the results are poor.

## Spray Can Application Tricks

Now that the various paints, dyes, and markers have been covered a brief mention to application techniques needs to be discussed. It is possible to mask out parts to control where a spray can or air brush deposits paint, dye, or marker. Simply cut painter's tape into long thin strips and use these strips to mask out regions you don't want to paint. Waterslide decals can also be used to mask and protect regions from paint, dye, or marker, check out the helmet used for the custom minifigure of Carly Foulkes. If using paint, just remember to use primer and follow the final color application with a clear gloss to give the final part a shine that resembles the original plastic.

For the legs it was much simpler, I masked out the legs, but not the shoes or the skirt. I did mask out the connection spots between the hips and the skirt as I wanted the figure to bend normally. After masking the figure I spray painted, first with a primer, as I don't want the figure color or the printing on the figure to affect the final color. After allowing the primer to dry I sprayed with a bright green spray paint. This results in an even and smooth finished to the figure. After this coat has dried I clear coated with a gloss protective finish. The *Friends* figure is now starting to look like Tinker Bell. This same masking technique could be used on any other LEGO element to alter the color of a small region.

With many options for altering LEGO part colors and you can create several new and different custom figures. Just remember, with all these options, it is best to figure out what works for the project at hand. If you need permanent hard wearing color alteration, look to your dyes, and if you need quick small area coverage look to paint or markers. Enjoy the new rainbow of colors now open to your LEGO elements and creations.

Custom figures aren't complete without that special accessory. After all, who is Superman without his cape, Batman without his cowl, or the Tick without his antenna? The most challenging aspect of the hobby can be finding that one accessory to complete the look required, sometimes that accessory simply doesn't exist. This means you have to construct that custom accessory. That custom accessory could be a hairpiece, cape or cloth element, weapon, tool, or most anything (i.e., is Doctor Who complete without his TARDIS?). As such, this chapter has to deal with multiple mediums (cloth, clay, styrene, resin plastic, and many more). Because of the diversity of this chapter I recommend practicing with one section or one material at a time. In an attempt to help the reader develop and build skills I have tried to place the contents of this chapter in order of complexity where possible.

*The Tick.*

*Amy and the Doctor in front of the TARDIS.*

## Cloth, Paper, and Film

Cloth, paper, or plastic film (for those of you old enough, think transparency film) can be added to a figure to make outer wear (Capes, Hoods, Skirts, Pauldrons, etc.) for the figure. These materials can be used to create the same types of enveloping accessories. By enveloping, I mean that these accessories typically wrap around the figure or some body part of the figure. This chapter will give you the basics info on how to create these types of custom accessories, which you will find are only limited by your imagination. LEGO has over the last few years started to learn this as well, for years they only gave us capes (sails, flags, & tents), but recently they have made grass skirts, kamas, ponchos, pauldrons, collars, kilts, graduation gowns, and even car covers. This speaks to the variety that can be made and LEGO has only scratched the surface.

Since cloth, paper, or film are interchangeable for the most part, I will refer to the starting material as cloth from here on out, just know you could use any one of the three different materials. Also know that a printer can be used to apply decoration to all of these materials. Cloth takes some special treatments, which will be covered towards the end of this chapter.

The first example I am going to show is Adam & Eve. Where would they be without their fig leaf loin cloths? The figures are very basic, detailing is minimal, without the fig leafs these would just be plain figures, with the cloth, apple, and snake their facial expressions take on new meaning creating that complete thought (content) and thus making an artistic representation of Adam & Eve.

*Adam & Eve*

Let's briefly go through the basics, check the first book for more details. Just as you have created templates for creating decals, measure and create templates for your cloth. You can mock them up out of paper, just know the cloth will stretch a touch and paper will not. Also, if you are making a cloth template that is going to fold around a figure part, include some extra material for the draping affect in your measurements to ensure the cloth part fits correctly. To get you started here are three templates made from the LEGO cloth so you can add your own details.

Templates: A cape (left), a short cape (center), and a pauldron (right), displayed above for your use.

Some of the tools used.

These soldier designs wrap the entire torso adding printable cloth to the figures, made by Mark Parker.

With a few templates, let's discuss materials. One can use LEGO cloth, paper, fabric (printable or regular), plastic film, or even leather. All of these are good options and fairly easy to work with if you are patient. Using LEGO cloth is the easiest custom cloth piece to make as LEGO has already treated the material so it won't fray. Fraying is the unraveling of the material at the edges; however, even LEGO cloth will fray if played with enough.

## Tools

The most important tool for this work is a sharp pair of scissors. Many seamstresses swear that once you cut paper with scissors they will never cut fabric the same, so they keep a set of scissors for paper and one for fabric. If you are a younger reader ask for help from your parents. I recommend Fiskas' brand scissors, which are a bit more expensive, but are worth every penny. You will also need an inexpensive leather punch tool with a rotating punch size (see the toolbox chapter page 14). Alternatively a set of wad punches in a few basic sizes (2mm, 4mm and 5mm).

## Anti-Fraying

You have to stop it without making your cloth so thick and stiff that it does not work/bend/fold well. One of the issues with making complex cloth accessories is that the cloth can become visibly thick and interfere with the arm or waist studs. There are a number of different methods and products you can use to stop your cloth from fraying. These range from diluted White (PVA – Elmer's) Glue to Acrylic Medium or even artists Acrylic Varnish. Treating a small swatch of cloth is typically easier than a large piece; just make sure it is big enough to get a few pieces per swatch (around 15cm by 25cm). Both the PVA glue and the acrylic medium can go milky on some fabrics, especially dark colored fabric. Try using artist's Acrylic Varnish, which is very easily applied with a small hobby roller and doesn't require any thinning. One major advantage of using the varnish is that you can paint it on any colored fabric and it will dry clear. So if you are able to get cloth that matches the LEGO color you want, it can easily be treated with the varnish, which can then be used to cut out your design. Keep in mind that adding any kind of treatment solution will make the fabric darker. Another good thing about the varnish is that you can actually mix it with acrylic paints to both color and stiffen your cloth. This trick works great for unusual colors! Just make sure you follow the directions on the bottle.

Remember that cloth isn't limited to minifigures, you can make flags, banners, sails, and even animal accessories, using the techniques outlined above. Just get creative, make new items and really customize your figures to the extent you can.

## Printable Fabric

Printable cloth is readily available at hobby and fabric stores in a variety of different materials. LEGO cloth is a type of broad cloth, so look for something similar (printable cotton). Printable cloth works much the same way as paper, feed it into your printer and add details. The advantage of printable cloth is you get to print the designs and the background color, so you can make your cloth part any color you want. Printable cloth is typically treated with an anti-fraying solution, but read the instructions that come with your package to make sure. If you need to add an anti-fray agent, stay tuned those details will follow shortly. Customizers in the know favor the Jacquard's printable cloth brand, but most any brand will work. If you cannot find printable fabric locally, check online. Some brands of printable fabric only allow you to print on one side, so if you are wanting the cloth to be the same color on both sides you might have to pull out your hobby acrylic paints, which work well on most cloths, including LEGO cloth. If you want to make it all you can even make your own printable cloth. By constructing your own cloth you can control all the colors of the cloth and even experiment with different base cloth colors. Just remember your printer, it doesn't have white ink, so a base color can be a light shade of most any color and you would print the darker shades on top. Home printers can't do the opposite; they can't print lighter shades on darker material, so keep this in mind when designing your custom cloth art.

There are many websites that have methods for creating printable fabric. The instructions that will follow include the most common options. See the tips at the end of this section.

*Adam and Eve feature custom printed fig cloths. Designed by Jared Burks and printed by Mark Parker*

# Creating your own Inkjet Printable Fabric

*Disclaimer: Fabric will be placed on a stabilizing material which will then be fed into your printer. Exercise caution and care, especially if the fabric gets jammed in the printer. This could damage your printer permanently.*

NOTE: Not all inkjet ink is equal and the type you use is the key to quality results. Cheaper printer cartridges and refills typically use dye-based ink, which will give unpredictable results. These inks may not be colorfast either, meaning they may wash out of your fabric. While it is unlikely you will wash your custom LEGO cloth part it is still something to consider. Better printers use pigment ink, which is typically colorfast on many surfaces (NOT DECAL FILM) and is better for printing on fabric. Identifying which type of ink your printer uses can be determined with a physical examination of the ink. When replacing the yellow ink cartridge remove some of the yellow ink and place it on a piece of clear glass. If your printer has pigment inks this droplet of ink will appear vibrant and opaque (meaning light does not transmit though the ink). If your printer uses dye based ink this droplet will appear dark dull yellow brown and be transparent. For improved results using dye based inks, Bubble Jet Set and Bubble Jet Set Rinse can be used, be sure to follow all instructions that come with the solution. Bubble Jet Set is available at Wal-mart and a simple Google search will turn up loads of other suppliers.

*Painter by Mark Parker*

# Materials Needed

Cloth (fine weave broad cloth – 100% cotton)
Freezer Paper - **Reynolds Freezer Paper**
Scissors
Iron
Inkjet Printer (with pigment inks preferably)

# Protocol

1. Cut a piece of fabric slightly larger than 8.5" x 11" (letter size – or whatever your printer is capable of printing)

2. Cut a piece of freezer paper slightly larger. Freezer paper is also known as butcher paper. It has a wax coated side and an uncoated paper side.

3. Place the fabric down on an ironing board followed by the freezer paper (coated side down). The working surface of the cloth is down and the uncoated paper side of the freezer paper is up.

4. Iron the two sheets together allowing the coating to fuse the fabric with the freezer paper. Do not use the steam setting on the iron.

5. Trim the combined fabric paper to 8.5" x 11" size.

6. Place this supported fabric paper into the printer, take note of the direction paper should be placed in the print tray placing the cloth side in required direction for printing.

7. Print, be sure to use high resolution images (300 dpi or greater) as low resolution could give pixilated results.

8. WAIT for the ink to dry (preferably overnight, especially for dye based inks).

9. Trim to desired shape and remove the freezer paper from the back of the fabric. Depending on the size and scale of your item these two steps might need to be reversed, peel and then trim, find what works for you.

## Tips

To prevent fabric jamming in the printer:

• CUT, don't pull, loose threads from the fabric sheets before inserting in the printer.

• Make sure the leading edge is securely bonded to the freezer paper. If the there is a separation this could cause a jam. You can apply a small run of tape over the leading edge, folded front and back, to help with the feeding of the fabric.

• Separations can occur because of wrinkles and they commonly appear as bubbles in the fabric. Load only smooth fabric sheets that are free of bubbles and wrinkles.

• Always practice printing with a piece of plain paper before using fabric sheets.

• Only add one and only one fabric sheet to the tray at a time to avoid jams.

• If a jam occurs never try to remove it from the printer until the printer is completely powered down. Make sure you have released all spring loaded rollers to allow the jammed fabric to be more easily removed. Follow the instructions in your printer's operation manual to identify and aid in jam removal.

• If a jam occurs and you clear it, get a flashlight and check the printer for any loose threads that may have separated from the fabric sheet.

## Modification

It is time to learn how to accessorize that figure. There are many different types of minifigure accessories and several ways to create them, most involve cutting, gluing, and mixing media. This section will discuss creating new elements through modification. Don't limit your imagination, get out that X-acto knife and pull out some LEGO elements; it really is ok to cut them!

This chapter deals with handling hobby knifes, hobby saws, and razor blades; all of which are sharp. If you are a younger reader please seek your parents' assistance in handling these items. If you are an older reader please use care and caution; your fingers are not replaceable. Kevlar gloves are available to protect your fingers, I recommend these to all. They can be found in most woodworking/carving stores or online from hobby sites. I also recommend a non-slip cutting mat as well as good technique. The best lesson I have learned is to use sandpaper when possible as many items can be created by sanding, more on this in a moment.

## Proper Cutting Technique

1. Always use adequate lighting.

2. Never cut towards your fingers, always cut away to avoid blade slips (Where the blade slips and can accidentally cut you).

3. Avoid holding parts in your hand while cutting. If you must hand hold a part to cut it, always wear a Kevlar glove.

4. If the part is thicker than 1/8 of an inch (minifigure handle thickness) use a hobby saw or Dremel Tool. These are much safer when cutting through thicker materials.

5. Use a hobby cutting mat and a desk or table, never cut parts in your lap or on odd or uneven surfaces.

6. Only use sharp tools, dull tools can hang and cause slips and accidents more frequently than sharp tools.

126. Tools of the Trade: Razor blades, hobby knives, hobby saws, sandpaper, Dremel tool (not pictured), cutting mats, Brasso, superglue, ruler, and protective equipment.

Check out some of the basic modifications in *Minifigure Customization: Populate Your World* to find out how to make weapons modifications. What will follow will be a bit more advanced than what was presented there.

## Hat Head

One of the greatest challenges with most custom figures is finding an appropriate hair style. In the last few years LEGO has started to create more hair styles that are more modern, especially in the Collectible Minifigure series. These are great, but as we develop as customizers the hairpiece is merely something else we can customize. The simplest alterations can be to change the color of the hair or make small adjustments to the shape. Existing hairpieces can be easily modified with sandpaper to create very different looking hair styles. A few examples will follow starting from easy to complex.

In the figures above simple color alterations are achieved with spray paint (be sure to use a primer, details in the paint section, page 29). Small alterations in shape can also be easily achieved with sandpaper or even a hobby knife. Be sure to polish the sanded or cut element back to a high shine using Brasso or high grit sandpaper. As mentioned in Minifigure Customization: Populate Your World, hair can also be combined with hats or helmets to produce interesting results. LEGO has started to incorporate hats and hair in several recent figures (including: Pirates of the Caribbean Jack Sparrow and Angelica; Collectible Minifigure S7 Viking Woman, S8 Cowgirl, and S9 Fortune Teller; Friends Riding Helmet; and the Lone Ranger's Tonto). As LEGO hasn't fixed all the "Hat Head" hair issues we can fix the ones we need with a simple technique of cutting the hair to fit the helmet and glue the two parts together.

Small alterations can be achieved with an X-acto knife, however to grossly reshape or shorten the hairpiece to create something entirely new is a more complex process. When we use an

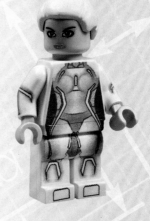

*Helmet hair has been previously examined, as has part color alterations using paint. The final example is a very slight modification to the front of the hairpiece to give the shape from the film. The back was also slightly modified.*

existing part to start with we are limited to what is present. Essentially we are performing the art of subtractive sculpting (remember this concept, we will discuss it in more detail in a later chapter), simply removing the undesired portions, i.e. giving the part a haircut! The key point to learn is how to alter a LEGO hairpiece safely.

Many hairpieces can easily be shortened, this includes helmets as well. This can be performed by merely removing the length or excess. I recommend this be done using a rotary tool or razor saw if the amount to remove is larger than 1/8 inch or 3 mm. If it is less than 1/8th inch or 3 mm I would merely sand away the undesired portion by hand. Just remember that once the gross removal is completed you will need to sand the part back to the high shine that the LEGO hairpiece is produced with, so leave a touch of excess that will be removed during this sanding step.

## Hair Shortening/Helmet Modification Steps

**1.** Gross removal of unwanted part. When you cut a part it becomes square. When a rotary tool is used to remove an unwanted portion of plastic it can partially melt if the tool is spinning too fast for the sandpaper or cutter to remove the unwanted plastic (i.e., they become clogged by the speed). So work slowly with a rotary tool at a slow speed to minimize this issue. The slower speed will be safer for you as the cutter is cutting the element which will reduce the chances of it being loaded or bound, lowering the possibility of an accident occurring.

**2.** Scraping/sculpting the part to shape. With the gross removal completed the part will need to be shaped back into a more organic or natural shape. Hair doesn't have squared edges in hair design. To fix this issue, using a razor blade, I scrape along the edge to round it back to a more natural shape. Also make sure the hair doesn't all end in a perfect line either. Imperfections make the style better.

**3.** Once the organic shape is achieved it is time to start adding/correcting details. When the gross removal was performed the edges and details of the hair were also removed or changed. It is time to add these groves or edges back to the hair to make it look more real and less plastic. This can be done gently with a small metal file or with GREAT care an X-acto knife.

**4.** Finally the part needs to be cleaned up to make it look like it is fresh out of the mold. This means either steps through sandpapers like the ones from Micro-mesh or a large amount of Brasso. Either will work, but the Brasso will take more time. This is because it is a constant grit, unlike the sandpapers, which will vary in grit and allow more broad removal at the beginning. If you take your time and make sure the lower grits are well done the final few papers will go quickly and result in a highly polished final result.

The next type of alteration is the removal of a particular portion of the hairpiece, specifically bangs, ponytails, etc. This is a much more delicate procedure to perform. This means that removal has to be exact. It can not merely be a guess. When I do this sort of removal I will not use a rotary tool as it risks too much and can damage the remainder of the part too quickly.

*Before and After: Shorting hair can be easy, just use care when holding the part. Use a razor saw when possible, rotary tool with a sanding attachment, or sandpaper when possible.*

## Hair Alteration Steps

**1.** Using a VERY sharp razor blade slowly cut away the undesired portion of the hair. USE caution and a Kevlar glove, especially if you are hand holding the part.

**2.** Once the undesired portion is removed examine how you are going to structure the hair that was under the undesired portion.

**3.** This typically means that the hair will have to be restructured/retextured. The areas affected are typically in prominent locations so this has to be done to tie into the existing behind and make sense of the area recently uncovered. Retexturing the hair will require grooves to be cut into the hair. I prefer to do this with a V-shaped or rounded-shaped metal file.

**4.** With the grooves cut into the hairpiece it is time to polish and sand the part back to its final state. This is again done using Brasso or Micro-mesh sandpaper. You can wrap the sandpaper around the small files to sand the areas in the grooves.

Alterations to hair doesn't have to be limited to what is on top of the characters head, beards can also be altered. In the example below a goatee is created by removing the excess from the beard. Recall above the explanation of an organic shape to hair, this is the exception. Beards are commonly trimmed such that there is a hard or sharp edge; the figure created below is the perfect example of this style of goatee. Also note that the sides of the beard were removed and that the connection to the minifigure was minimized to give it a much more subtle appearance.

## Advanced Modification, Basic Sculpting, & Basic Molding

Many things are possible and they don't always start with creating a brand new part from scratch. The previous section focused on part modification creation via the art of subtractive sculpting. Let's think about sculpting through addition (remember this concept, we will discuss it in more detail in a later chapter) or what we can create with the addition of clay (this is a more advance method). Therefore, the original hair style must be considered. The advantage is that the resulting part is typically more durable as it has a plastic base.

With this in mind the next example mixes part modification with clay addition, basic sculpting, and basic molding and casting (Sculpting, Molding, and Casting will be discussed in detail in later sections, just remember this foundation). In short, a hair style to mimic that of Tinker Bell was needed. LEGO has given the foundation for a Tinker Bell Hair Style in the Series 8 Fairy and the Top Knot Bun (these hairs elements are found in the collectible Minifigures Sumo Wrestler and Kimono Girl). Since the Top Knot Bun has minimal details, the Fairy Hair was used (Top Knot Bun and Forelock).

## Advanced Hair Alteration Steps

**1.** Using a rotary tool the Forelock (bangs) was sanded off of the front of the hairpiece.

**2.** Clay was then added to the front to make the hairpiece look more like Tinker Bell's hair.

**3.** Several flowing bangs were sculpted into the hair to make it look a bit windswept; she does fly around all the time after all.

*Before and After: Removing hair from the front can be done, it just takes some trial and error. Work slowly to ensure you don't remove too much and be sure to add back the details as seen in the close-up.*

*Before and After: Beard shortening is also possible. Just remember this is an exception to every rule, as many beard styles trim to an exact edge. Thus with a quick cut the beard above was shortened. The sides were also removed as this is more of a thick goatee than a full beard.*

*Adding bangs using clay for Tinker Bell.*

*Before, during and after adding bangs.*

*Ear transplant.*

**4.** Clay was cured in toaster oven at $175^0$ F for 20 minutes.

**5.** Ears were clay molded.

**6.** Ears were cast.

**7.** Ears were added to Tinker Bell Hair style.

**8.** New Hair Style was cured with clay ears in toaster oven at $175^0$ F for 20 minutes.

**9.** Either Hair could be painted OR Hair could be molded and cast with pigmented resin.

Basically the bangs from the S8 Fairy were removed and new ones were sculpted in their place. After the bangs were correct, pictures of Tinker Bell were examined. She is a Fairy with pointy ears, which could be created by sculpting; however the desire is to make them resemble other LEGO ears. Therefore, LEGO hairpieces from mythological figures were examined specifically from the Lord of the Rings and Collectible Minifigure Themes. Orc and Cyclops ears were either too large or the wrong shape. Elf/Werewolf ears were perfect, but how to get the ear off one of the Elf/Werewolf hairpieces and on to Tink's. This is a simple trick, an ear mold is needed.

## Ear Mold Creation

**1.** Take some clay, mash it into the part you want to mimic, gently remove it from the part, and cure in the oven (toaster oven $175^0$ F for 30 minutes). You will need 2 molds, one for each ear.

**2.** Once cured and the heat has dissipated simply take some clay and mash it into this tiny mold, remove the excess, and then ever so gently remove the uncured clay from the mold.

**3.** Add to Tink's hair.

**4.** Adding the ear with this technique will cause it to stick out a bit in relation to the hair, so it will have to be sanded down after cured.

**5.** Paint or Mold and Cast with pigmented resin.

## Tips & Tricks

After the part is completely created give it a light spray with a high gloss paint to fill any imperfections so that the final part is smoother. This will make the final part more shinny, be it painted or molded and cast.

Advanced modifications can come in multiple forms; the one above is just one example. As mentioned before, modifications are a form of mixed media art. Mixing mediums refers to having two or more different substances (or mediums) coming together to create something new. Be sure to explore every medium you can find, they all bring something different. While this book will present mixing clay, various plastics (each with their own properties), epoxy, and a few other materials; be sure to explore. You might just find the next best thing to create custom accessories.

## General Modifications Tips & Tricks: Adhesives

Almost all modifications involve using some form of adhesive to attach one part to another. Be sure to understand the two materials that you are trying to adhere together. Also note that many adhesives can be toxic and shouldn't be used in enclosed, confined spaces. As a default adhesive I reach for cyanoacrylate (super glue).

It has a strong bond and sets very quickly. I prefer the gel variety for the added control. If I am bonding (welding) to pieces of plastic together where I want to disguise the joint between the two pieces of plastic I will commonly use Plastruct Plastic Weld. This is a solvent and it will surface melt two pieces of plastic together. Recently I have found (on the advice of a chemical engineer) another solvent adhesive called "Same Stuff" from Micromark. Same Stuff features a capillary delivery system of the solvent, which more precisely controls the placement and yields better results. Make sure your adhesive works for your application.

# Sculpting

Sculpting has been discussed in two different forms so far in this chapter, subtractive and additive. As their names suggest, one can create items by either removing or adding material, it really all depends on the creator, find a style that works for you. Generally I sculpt through addition, this allows me to sand and surface all the parts as I go to get the final part exactly as I want it with the desired finish. There are some tricks to achieving the best results. In this section clay versus wax sculpting will be discussed as well as proto-casting. Before we dive into these details we need to discuss the basics.

## The Basics

### Types of Media: Polymer Clay, Epoxy Putty, & Wax

There are several types of clay ranging from the earthen clay dug from the ground to completely synthetic, most are inappropriate for this type of application. The major type used for making custom pieces in LEGO scale is polymer clay, however, some use Epoxy putty (MagicSculp) which is similar to clay. Polymer clays will not harden till cured by low heat; epoxy putties are cured by a chemical reaction that is started when the two parts of the putties are mixed together. This typically limits your sculpting time to less than 24 hours. Choosing one over the other is ultimately up to your preference, however there are some basic arguments for both. It boils down to extended sculpting time versus durability. The clay will not cure till heated and the epoxy putty will be more hard wearing. If you are not going to mold and cast your part and/ or you are a fast sculptor I recommend the epoxy putties, if not polymer clays.

Polymer clay has a relatively low curing temperature and remains pliable until cured. Polymer clay hardens by curing at 265 to 275 °F (129 to 135 °C) for 15 minutes per 1/4" (6 mm) of thickness in your home oven and does not shrink or change texture during the process. The curing temperature can be lowered if the clay is heated for a longer duration (this is particularly helpful if you are sculpting on top of an ABS plastic part, which has a melting temperature of 221 °F (105 °C). You can also cure polymer clay by placing it in very hot water or surface cure it using a hair dryer. Surface curing will allow you to sand your part; however without a complete curing the clay can be fragile. When properly cured polymer clays are quite strong and won't normally break when stressed or dropped. Polymer clay is available in hobby and craft stores, and even stores such as Wal-mart. Leading brands of polymer clay include Premo, Fimo, Kato Polyclay, and Sculpey. These clays are available in a wide array of colors so you might not even need to paint the parts you create. On top of these great options this type of clay is quite inexpensive; a small package of clay, more than enough to create many parts and can be purchased for less than $2. I prefer

*Sculpey Firm & Magic Sculp*

*Commercial clay tools.*

Sculpey firm, which is as its name suggests is firmer clay. This firmness makes it a bit easier to handle for me.

Epoxy Putty has many of the exact same characteristics as polymer clay with the main difference being the mechanism of curing. Epoxy putties are mixtures of two chemically different parts. Mixing the two parts together initiate the chemical process of curing the putty. This curing time depends on the thickness of the putty and how well the two parts were mixed together. Typically a 24 hours cure time is common for a minifigure scale item. The big advantage is the durability of the epoxy putty as it will always completely cure.

Wax, pardon the pun, is a whole different ball altogether. Wax is a vague term for this item; it is actually a mixture of several different types of wax. The composition is what makes it hard or soft, and it affects the level of detail achievable. Wax is typically used as a secondary product for sculpting, meaning that the artist starts in clay and works to a basic to intermediate detail level. Then the clay part is molded and a wax part is cast from the mold (wax is melted and added to the mold and removed once solidified). This cast wax part can now be sculpted to achieve a high degree of detail. The wax piece is fragile. It will ding if dropped and like all wax it is sensitive to heat. Therefore once created it is molded and cast to make plastic versions.

In this section we have discussed three different sculpting materials, each with pros and cons, for uniformity from here on out I will refer to sculpting material as clay. Once you get your hands on clay (putty or wax), play with it, work it, and get a general feel for it. Your hands are your best tools so play with some of the material and see what level of detail you can capture with just your hands, don't forget your fingernails, they can be used to sculpt details as well. You will likely need some additional tools to create the fine details on your creations, and maybe even a texture mold (similar to the ear mold previously discussed). I find any fine tipped item works well including X-acto knives, paper clips, and most any other small item. Find items that work for you. Your tools don't have to be store bought clay tools; most of these are actually too large for this work. If you want buy a tool set, look for dental tools (Harbor Freight has some good options). Remember this is very small scale work; your tools need to be able to create fine detail. The best advice I can give you on tools is one I recently picked up and is a modification of the trick presented in the first book, sculpt using chopsticks. Basically take one of the chopsticks and sharpen the point in a pencil sharpener or cut the tip to most any shape you desire (look at larger sculpting tools and replicate their shapes on the chopsticks). Now you have a customized sculpting tool that cost you nothing (unless you bought the chopsticks specifically for this purpose, then it likely cost you pennies). If you lose it or if it gets damaged, simply make a new one.

## Sculpting 101

As mentioned, sculpting is created by addition or subtraction of material. One tip I can offer is subtraction works better with cured clay, putty, and wax. However it is possible to add material to any of sculpting options. With clay or putty this is simply pushed onto the existing material and worked on to bond. With wax it is a touch more difficult. A tool called a waxer has to be used. Essentially this is a heated wire that is used to transfer melted wax to the object you are sculpting. You also have to use this item to surface melt the item you are sculpting so the new wax and old wax bond together, or the new material will simply flake off. Additive sculpting is a very useful approach and is also referred to as layered sculpting. Layered sculpting is a technique where lower layers can be sanded/polished before upper layers are added. When the upper or top layers are added it is not always possible to sand/polish the lower layers completely because they are obstructed by the upper layers.

Unfortunately, I cannot tell you exactly how to sculpt a part, it will take time and practice to master this art. I can offer a few tips and tricks that I have learned to help speed your learning curve into creating a custom piece. I can also show you a few examples for items I have created.

## Design Suggestions for Creating Custom Hair

If you want to sculpt a hairpiece that works in the LEGOverse avoid too much detail as LEGO commonly stays simple by designing hairpieces that display locks or segments of hair as opposed to individual hair strands. Make sure your design frames the face of the minifigure as your figure wants to see who is checking out their new hair style. Covering the face is not a good idea. For a more realistic hair style avoid perfect symmetry in the design style. Make sure that the hairline starts at the curved portion of the head or slightly lower.

A very helpful trick I have learned is the creation of a texture mold; this is the same concept as the ear mold shown earlier. If you can identify an official hairpiece or any other piece of textured plastic that has a similar detail to the hair style you wish to create you can press clay onto that piece of plastic and cure the clay. This original piece is called the "master." Once cured the clay can be removed from the "master." The clay now functions as a one piece mold, which can be pressed into the clay on the hairpiece you are attempting to create to translate the texture to the hairpiece you are creating. You will still have to sculpt the hair into the desired general shape, but this will add the detail. These details will need to be cleaned up slightly, but this method will add detail quickly and your press mold can be altered in direction and angle to create very different details to a hairpiece.

Alternatively, when sculpting all the details from scratch I create my parts in layers. This allows me to sculpt and sand as I go, when I am finished sculpting I have little sanding left. This is exceptionally critical with small parts like hairpieces. If you try and sand the part at the end, you may find that the detail work on the piece is difficult to work around. Sanding is a slow process, but it is CRITICAL to making your custom element look more like LEGO elements. Two tips to save time on your sanding is right before you cure (surface or complete) any part; give the part a quick wipe down with a cotton swab or paint brush that has been dipped in rubbing alcohol (70% Isopropyl Alcohol) for clay or water for epoxy putty. This will remove any fingerprints or other slight imperfections on the sculpted part leaving a smoother surface. The second tip is to scrape the area you want sanded with a very sharp X-acto knife. After scraping or wiping down with alcohol or water it is time to sand. To sand your custom parts you will need very fine grit sandpaper, which can be purchased at most hobby stores (woodworking grades are simply too coarse). Because you are using an ultra fine grit paper, sanding will be slow, but you will be rewarded in the end with a better looking piece. As with the tools you are using

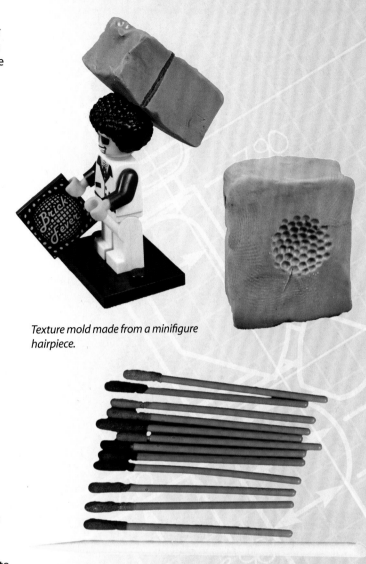

Texture mold made from a minifigure hairpiece.

Sanding Tools. Modeling sanding sticks from Hobby Lobby (blue & white). Micromesh sanding swabs.

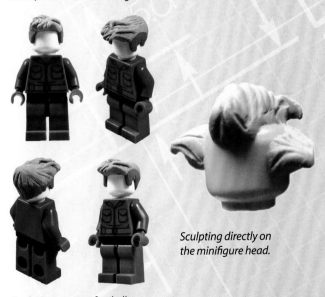

Sculpting directly on the minifigure head.

Sculpting on top of a skullcap.

*Finished Figures: Bubbles (top) and Honky from the Clown Commandos.*

to sculpt the part you will need small tools to sand it. To make these sanding tools cut the sandpaper and attach it to small sticks or rods. Make sure to step up in grit values to really get the best finish. Another trick to help with the finish is to paint the final product with a HIGH gloss spray paint. This will fill small imperfections and give a shiny finish to the part.

## Tips and Tricks

If you are creating a hairpiece or some other accessory item that must attach to a LEGO element, you will likely want to remove it after you have finished sculpting it. This can be tricky and if your clay isn't completely cured you could damage it by trying to remove it. To help with part removal, you can wrap the LEGO element it attaches to with very thin aluminum foil or parafilm. Parafilm is a stretchable wax like product that is used by hobbyist when painting models. It can be found at hobby shops or online. Another great trick is to sculpt over the top of another element. You can sand down a LEGO hairpiece to a "skullcap" and add clay to the top of it allowing you to sculpt a new hair style, helmet or whatever. Most importantly this allows you to keep the internal stud acceptor and more easily remove the cured part. Just remember when curing this clay to be careful, you will need to cure using slower/lower temperature technique previously discussed. If you plan on molding and casting your finished piece merely include the head in the molding process.

## The Wrap

I will end this section with one piece of advice; sculpt, sculpt, and sculpt. Only through practice will you get better. Sculpting and re-sculpting a piece will teach you something each time you make the design. Everyone needs a hair dryer, even if you are bald. Sandpaper is your friend, sand early and often, always increasing in grit. Seal your final part to give it that LEGO sheen.

## Tips and Tricks

Depending on what you are creating, especially if the item is long and thin (swords, etc); your creation will need an internal support skeleton. The support can be made of most anything. I have used everything from wire to wood, but I prefer styrene. Styrene is a type of plastic which can be purchased in sheets and a variety of shapes from most hobby stores. By using an internal support skeleton your item will be stronger. The reason I prefer styrene is it has a slightly higher melting temperature than the curing temperature of polymer clay. This doesn't mean you can merely pop items containing styrene in the oven. However the styrene is reasonably stable when the polymer clay is cured by boiling, hair dryer, or extended time low temperature curing methods. Styrene/plastic can be incorporated into your sculpture as well. Certain shapes are very difficult to sculpt perfectly. Another tip is if you are creating pieces with a grip (sword hilt, for example); make the grip portion out of plastic. This will help make your part more durable as the plastic will take the abuse of the part removal from the figure's hand, protecting the clay. Be sure to add a pin in the plastic portion of the handle. Add clay around the pin to extend the interacting surfaces between the two parts, strengthening the bond between the two pieces.

## Wax Sculpting

Just a quick note on wax sculpting, generally one does not start from a wax form or wax block to sculpt an item. As mentioned above wax is generally cast from a molded clay sculpted item to allow higher level of detail to be created. This is what I refer to as prototype casting (protocasting), which will be discussed after molding and casting. Just know that many pieces in LEGO form do not require the level of detail that wax can create, so be selective with its use.

## Pressure Molding & Casting

In *Minifigure Customization: Populate Your World,* the concept of molding and casting was presented. Mold and casting were discussed to demonstrate how one could take the clay sculpted part and duplicate it easily in a more durable material. Now it is time to show how to make this technique a bit easier by using pressure. When adding pressure, safety is always a concern as you are going to be stressing the container you are using (more on this shortly). As mentioned in the first book the one issue that plaques moldings and castings on this scale is air bubbles (trapped or formed). Air bubbles are caused by either moisture in the resin (forming via foaming) or the small thin regions just being difficult to penetrate by the resin (trapped). Pressure can assist with both these issues, by raising the relative pressure bubbles will either be crushed or they simply can't form.

Thinking back to the info on the 3 basic states of matter (solids, liquids, and gasses), the only easily compressible state of matter is gas, liquids and solids are not easily compressed. Since gas can be easily compressed the air bubble issue can be easily resolved through the use of pressure. Because we are going to pressure cast it is important to pressure mold to ensure that the mold doesn't contain any trapped air bubbles. In order to pressure mold and cast there is some additional equipment needed, specifically a pressure pot, air hoses, air compressor, and safety valves. The basic concept is that the mold box filled with silicon or the mold filled with resin is placed into the pressure pot, which is then sealed and the air inside is compressed to ~40 psi. When compressing the contents to 40 psi it is sufficient to crush and inhibit most bubble formation.

Before continuing safety must be discussed. When using pressure all safety options must be employed and you must regularly inspect your equipment. Follow all manufacturer recommendations for use and care of your air compressor and pressure pot. Make sure to inspect the air pressure release valves to make sure they are fully functioning. These values bleed off the pressure if it becomes too great to keep the pressure pot from exploding and are critical for your safety. Yes, I just use the term exploding. When air compressors and pressure pots are used safely this is not a likely occurrence. Also note all pressure pots have a maximum safe operating pressure, this value should NEVER be exceeded. Because of the nature of pressure pots, I always recommend purchasing a pot capable of handling the pressure, not trying to construct one on your own. Honestly, trying to find a pot capable of twice the desired pressure gives added safety. There is an economical option from Harbor Freight called a paint pot that is used by commercial house painters, which can be easily used as a pressure pot (*http://www.harborfreight.com/2-1-2-half-gallon-pressure-paint-tank-66839.html*). This paint pot is capable of 80 psi and we only need to pressurize it to 40 psi, so this will be well within the safety margin.

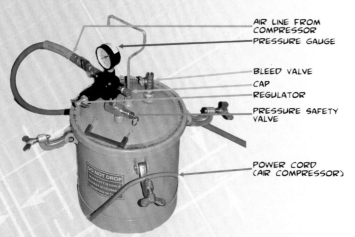

AIR LINE FROM COMPRESSOR

PRESSURE GAUGE

BLEED VALVE

CAP

REGULATOR

PRESSURE SAFETY VALVE

POWER CORD (AIR COMPRESSOR)

*Schematic for Pressure Pot.*

*Pressure Casting Results, no bubbles!!!*

*Prototype: Custom prototype element for molding created with sheet styrene.*

You will need to add a few parts to convert the paint pot to use as a pressure pot. Basically the paint straw is removed and a quick fitting is placed in its place to allow the pressure line to attach.

Now that you have created a pressure pot whenever you mold or cast simply place your mold or item you are molding into the pot to cure under 40 psi of pressure. This will compress the air inside the mold keeping bubbles at bay and it will push the liquid further into the mold to help fill the voids. The other critical feature that pressure casting gives as a result of no internal bubbles is the ability to sand the parts. Because there are no bubbles the parts can be polished to a high shine without fear of revealing subsurface bubbles.

## Silicon Rubber Molds

This section will present information on creating silicon rubber molds, which is repeated from the Minifigure Customization: Populate Your World. The big change here is to do it under pressure. As you will be using chemicals in not only the creation, but also the use of silicon rubber molds, please READ all safety information for all the products you use. I highly suggest using platinum cure silicon rubber as they are commonly used in medical and food handling applications, as such they contain safe chemical components. The creation and use of silicon rubber molds by younger readers will require adult supervision.

The custom elements that you have created will now be referred to as prototype parts or prototypes. This is because they are one of a kind. As they are unique you may wish to replicate them or simply make them more durable. Silicon rubber molds are the best option to accomplish either of these means. There are many manufactures of silicon rubber for molding including; Smooth-On, Alumilite, MicroMark, and others too numerous to list (the options listed are the most accessible). When making a silicon rubber mold two compounds, silicon rubber and catalyst, are mixed in a proper ratio to start a chemical reaction causing the silicon rubber to cure to a solid state. Many brands mix by a weight ratio, meaning a large amount of silicon mixed with a tiny amount of catalyst. These weight ratio mixes require the use of a gram scale to accurately recreate this complex mix ratio. However, the Smooth-On brand features a convenient 1:1 mix ratio, eliminating the need for very accurate weight measurements as they mix by volume.

## Silicon Rubber Characteristics

Now that we understand the differences in mix ratios we need to discuss the differences in the many types of silicon products. These different types allow for the creation of molds with different features. Therefore it is important to understand what each type of silicon is for and it's weakness and strengths. There are four key features to understand about silicon mold rubber; elongation at break, tear strength, pot life, and demold time. The first two and last two characteristics are related.

Elongation at break is just as it sounds; how much will the silicon stretch before it breaks. This measure is typically presented as a percentage. A low percent stretch is about 250% and a high percent of stretch is 1000%. These percents are not arbitrary, low percent means the mold is more firm and high percent are less firm (the elongation at break is a direct reflection of the Shore A strength). Think of this like jelly versus Jell-O. Producing multiple part molds with a high percent stretch is very difficult as it will emphasize part lines in the mold, so look for higher Shore A values. Part lines occur where multiple part molds meet. These are emphasized as the edges of the mold curl slightly due to the firmness issues. Tear strength is an indication of the force required to tear the rubber, it is measured in pounds per linear inch (pli). The higher the tear strength value the stronger the rubber. Typically higher tear strengths go hand in hand with greater elongation at break percentages. So the same issues that plague multiple part molds with higher elongations at break do so with higher tear strengths silicon as well.

Pot life is the duration you have to mix the two reagents, silicon rubber and catalyst, together and pour the mixture into the mold box. The mold box is the container that holds your part while you are creating the mold. The shorter the pot life the faster the silicon starts to set. Typically 20-30 minute pot life silicon is preferred. This gives plenty of time to completely mix the silicon (which is key to achieving the proper elongation at break and tear strength) and to pour the silicon into the mold box in a **slow** and **controlled** manner. The demold time is the duration that it takes for the silicon to completely cure or set. It is important to wait this duration before disassembling the mold box and proceeding to the next step, be it molding a second part of the multiple part mold or using the mold to cast the a new part in resin.

Summary: Look for Silicon rubbers with at least a Shore A of 20 or greater, an elongation at break of 1000% or less, and a pot life of approximately 25 minutes and a demold time of 4 hours or more. If 3 or more part molds are going to be used, a higher shore A value is required, which means imbedding the prototype is more difficult as the mold will not stretch as much to release the finished part.

# Mold Design

Now that you have an understanding of the characteristics of the different types of silicon we can begin a discussion of mold creation. A properly designed and executed mold will yield many great parts; a poorly designed and/or executed mold might not yield even one. The mold design can also affect how much part clean up is needed after, spend the time while making the mold to save time on every part created.

The first issue to consider when creating a mold is how the cast element will be removed. Typically, removal is the factor that determines how many parts to the mold you will have; one, two, or many. It will also affect which silicon you use. Once you have decided this issue the other key factors can be consider; mold box size, holding the prototype while pouring the silicon, part line locations, properly designed air vents and fill holes, and pour speed and technique. Mold boxes are easy; we already have a perfect product to create a mold box of any size, LEGO bricks. It is best to allow 3/16 to ½ inch (~1.25 cm) or more of silicon to surround the prototype to give the mold the proper strength and rigidity. Therefore when laying out the mold box in LEGO bricks allow this much space around the prototype. Once your mold box is complete you need to measure its length, width, and height in centimeters (cm). When you multiple the three numbers together you will have a measurement in cubic centimeters which is equal to milliliters (mls). Milliliters is a volumetric measure, when you divide this number in ½ you will know how much of each part of the 2 part silicon to use.

The next issue is suspending the prototype in the box. Most

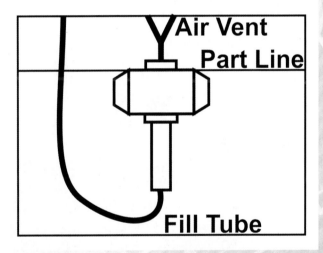

*Mold Design: Proper planning includes sketching a mold design to include part lines, air vents, and fill holes.*

molds for these types of prototypes are at least 2 parts. Therefore the prototype will be suspended and the first half of the mold will be created, then the first half will retain the prototype while the second half of the mold is created. Typically the prototype is embedded in clay to retain it while the mold is poured. Not just any clay, but a non-drying clay like Klean Klay. When the prototype is embedded in Klean Klay this will create a part line, the line formed between the two halves of the mold. So place the clay along a line in the prototype to hide the part line so it is less visible on the final cast element. Part lines on LEGO elements are visible under close scrutiny. The Klean Klay is also used to create keys in the two mold halves so they "lock" or register together. These are critical to proper mold alignment. Use isopropyl alcohol and a paint brush to brush along the edge of the part embedded in the Klean Klay to clean the separation line that will be created.

*Mold Box: Mold box set up with Klean Klay used for part suspension and lock and key design for mold registration.*

Properly designed air vents and fill holes for a mold are the most critical factors in creating a functional mold. Air vents and fill holes are where air trapped in the mold escapes as it is filled with the resin plastic. These need to be placed such that they can be trimmed and not affect the appearance of the finished cast element. Typically the mold is filled from its lowest spot and vented from its highest. Designing good molds take time and practice as well as 3-dimensional thinking. The way a prototype is held can affect the air vents and fill holes so consider this when embedding the prototype in Klean Klay. When I create a mold I use small diameter styrene rods to create the air vents and fill holes in the mold while it cures. When the mold has cured the rod is removed and the voids serve as vents and/or fill holes. Many suggest cutting these into the mold after it has been created, I do not recommend this as it creates irregular vents that can trap air. Air trapped in the part region of the mold when casting is the worst issue as it will result in a poorly created part. While pressure will assist with this issue, it can be complicated by irregular hand cut vents, another reason to cast them in with styrene rods.

## Pouring the Mold

The final item to consider when pouring a mold is actually mixing and pouring the silicon for the mold. When mixing the two parts, pour part 1 and 2 into a mixing container (I prefer paper or plastic cups) in the proper ratio. Mixing with a metal or glass rod will reduce the amount of air introduced into the silicon, which is ideal. Many molding and casting kits come with broad wooden sticks; these will whip in more air, which could cause problems. When completely mixed pour the mixture into a new container and continue to mix. Properly mixed silicon is critical in yielding a mold with the elongation at break, Shore A, and tear strength indicated by the packaging. Improper mixing can result in weak spots in the mold and result in a short mold life or poor part production. A properly mixed and cared for mold should be able to produce 25-200 parts, depending on the types of silicon and resin. Remember to keep an eye on the clock as the silicon will start to set as you approach the pot life duration and you have yet to pour the silicon into the mold box. No need to rush, typically you have 20+ minutes.

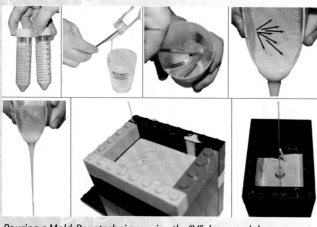

*Pouring a Mold: Pour technique using the "V" shape and slow pour rate to remove air whipped into the silicon resulting from mixing the two parts together.*

Once the silicon has been completely mixed it is time to pour it into the mold box. Pour technique can help reduce air bubbles in your mold, which can weaken and reduce the mold's functional life. Remember we are molding under pressure, so small bubbles aren't an issue, the air pressure will crush them. This takes some of the pressure off of good pouring technique. Just don't pour directly over your item you are trying to mold, allow the silicon to flow over it from the sides. That being said it never hurts to pour well. Pinch the edge of the cup forcing it to form a V shape. Then pour slowly in a thin stream, which will pop large air bubbles in the silicon. This is because the air bubbles are stretched as they fall out of the cup in a fine stream. Remember to allow 3/16 to 1/2 inch (~1.25 cm) or more of silicon on top and bottom of the part as well as to each of the sides. The amount on the top, where the air holes/fill holes are located is much more critical. If you don't have a good long fill line you can compress the air inside filling the voids from the fill lines, without sufficient resin in these lines you won't have a completely filled mold. Longer fill lines give a reservoir of resin to fill voids.

If your prototype requires a multiple part mold be sure to coat the area where the first and second mold parts meet with silicon to silicon mold release. If you do not the two mold portions will bond and you will NOT be able to separate them from one another. With proper application of the mold release (silicon to silicon, there are other types) the two parts will easily come apart allowing the prototype or cast element to be removed. When you are pouring the second part, exhibit care to not disturb or remove the prototype from the first half.

Following the above instructions you will be able to easily create a mold of one of your custom sculpted elements. Two part water thin plastic resins used in casting cure by an exothermic reaction (releasing heat). This heat slowly wears the mold. Be cautious not to cast too many elements at one time as you could lower the life of your mold. Let it cool between castings. To get the longest life from your mold it should be stored in a cool dry place. Happy Molding!

# Pressure Casting

This section is packed with details on casting resin plastics (typically urethane plastics) under pressure. Resin casting will be using chemicals in the creation of custom elements using silicon molds, please READ all safety information for all the products you use. The creation and use of resin plastic by younger readers will require adult supervision, especially with the use of pressure.

# Resin Characteristics

Resin plastics are composed of two parts much like the silicon rubber. By mixing part A with part B an exothermic (heat generating) chemical reaction takes place curing the resin into a hard plastic. Most resins are a 1:1 mixing, however a few are not, so read the instructions carefully for the resin you choose. I recommend the 1:1 mix for ease of use. Resin has a pot life, demold time, viscosity, tensile strength, and hardness, much like silicon rubber. These characteristics effect how long the resin will take to cure, how well it will flow in the mold, and how strong the final part will be.

Resins typically have a much shorter pot life and demold time than silicon rubbers. Because of this shorten duration resin must be quickly mixed and poured into a mold before it cure. If mixing takes too long the resin will start to cure and thicken keeping it from being pourable and thus it won't enter the mold. Most resins are referred to as water thin; this is in reference to their viscosity. A low viscosity resin will pour very easily. This means the resin will more easily fill small voids in the mold.

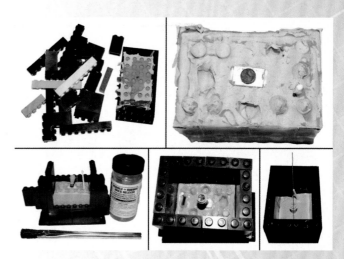

*Disassemble the mold box for Klean Klay clean up. Reassemble the mold box. Coat the top with rubber to rubber mold release and pour the second half of the mold.*

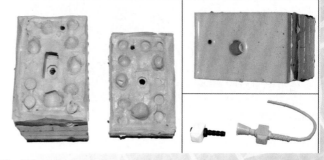

*Final Mold: The finished mold with original above and replica cast parts below.*

Tensile strength and hardness refer to the strength of properly mixed resin plastic. If the resin is improperly mixed or it contains bubbles the strength will suffer. The hardness also indicates how durable your final part will be; keep this in mind if you intend to sand the final product or if it is intended for rough play. Casting your plastic resin under pressure will help eliminate bubbles, allowing you to sand the final product to further improve or modify the surface if you so choose.

I prefer resins with 3-5 minute pot lives, anything shorter is hard to properly mix and get into the mold. As with silicon be sure to mix thoroughly; this means mix in one cup and then transfer to a new cup to continue mixing. My rule of thumb for mixing is one-third of the duration of the pot life. This gives you the other two-thirds of the pot life to pour the resin into the mold and get the resin filled mold into the pressure pot. Note improper mixing can also affect the cured resin color, especially if pigments are used.

## Coloring Resin

Most resins will turn white when cured, however there are some that turn; amber, clear, tan, opaque, and yellow. If you intend to color your resin you need to keep this characteristic in mind as it will affect the colors you can achieve. For example it is very difficult to achieve black colored resin when using white cure resin. When mixing the black pigments with the white cure resin the resin will turn light or dark grey depending on how much black pigment is added. One would have to use a clear or amber cure colored resin. It is possible to work around the color issue mentioned above for most colors. Typically adding pigments can have issues of foaming as you change the moisture content of the resin, however with the addition of pressure, many of these issues are resolved.

Most companies offer pigments in the following colors; white, black, red, blue, yellow, green, and brown (primary and secondary colors). These colors must be mixed together to achieve the color you desire. Creating custom color recipes will take quite a bit of time, so make sure to take notes. This does not mean you will have perfect color matching or consistency. Day to day humidity changes and mixing variability will affect the final color.

Some pigments are available in a powder form, which is great from a foaming stand point, but this makes them much more difficult to use. Measuring powders to determine the quantity to add to resin is difficult. Also powders have an ability to find their way into places they shouldn't. Remember these are VERY concentrated pigments so very little on your floor, clothes, or hands could result in a massive mess. If you choose to go this route be very careful and get a VERY accurate scale.

## Filling the Mold

Filling the mold should be easy and it can be with the right materials. As previously mentioned when the resin parts are mixed together an exothermic reaction occurs. This will ultimately destroy the mold so take every precaution to protect and extend the life of the mold with each casting. First, the mold should be sprayed with mold release, I prefer Mann's Ease Release 200. This will help keep the silicon rubber from drying out due to the heat and thus extend the life of the mold. This will also keep the mold from sticking to the resin once cured (if

*Pigments.*

*Mold release.*

using platinum cure silicon rubber molds this isn't as essential as resin will not stick to this type of rubber). Spray the mold with a thin even coating before each use and then use a dry paint brush to gently brush over the mold to make sure every surface is evenly covered. Allow the mold to sit for 3-5 minutes for the release agent to dry. After casting it is also best to let the mold rest and cool, the silicon will hold heat for some time and it is this heat over time that will destroy the mold. Immediate repeated castings can expedite the mold failure.

To fill the mold I prefer to use transfer pipettes. These are small plastic graduated tubes with a squeeze bulb on one end. This allows the user to uniformly push the resin into the mold. I find that this pressure helps to fill the mold very uniformly, if the mold is properly designed. This also avoids the mess of the thin stream pour and speeds the resin into the mold while still water thin. However, because this is not a thin stream pouring we risk the addition of bubbles. By casting under pressure we don't have to worry about this issue.

All resin volumes slightly shrink when curing. This means if you completely fill the mold when you take it apart the resin will have pulled back into the mold slightly. If the air/overflow tubes are short, this shrinkage could pull air into your mold, especially when casting under pressure. This issue can be counteracted with the addition of some small straws or reservoirs in the air/overflow areas if these regions were not properly designed in the molding process.

## Wrap-up

Now that you have the secrets to basic molding and casting under pressure let's see what you can create. Be sure to reread all the sections in this chapter as each will give small insights that can improve your work. This is a process and by cheating at any step you will sacrifice the results of the next. Enjoy making your own plastic parts.

## Proto-casting

Now that we have discussed pressure casting we can think about the concept of proto-casting. Proto-casting is similar to sculpting in wax, it allows for an intermediate part to be created that can then have additional details or higher degree of polish to be achieved. Typically when creating a part in clay the desired detail or finish can't be achieved. However, a quick single part mold where a part is molded around an existing LEGO part can be created quickly. This gives a quick, yet imperfect part that can be sanded and modified to make the part you are after because of the use of pressure removing all bubbles in even this proto-part. By using Micro-mesh sandpaper a polish greater than the finish LEGO has on the brick can be achieved. This can then be molded using platinum cure silicon rubber and this finish will be on all the subsequent part created from the secondary mold. So instead of thinking about molding and casting as a final step, it can also serve as a stepping stone.

## Vacuum Forming

The kids of the 1960s had it good! Yes, that statement sounds odd, but think about it for a second. In the early sixties LEGO came to America, cause enough to celebrate, but this isn't the only reason to think these kids had it good. The Mattel Corporation had a toy ahead of its time for this hobby, the

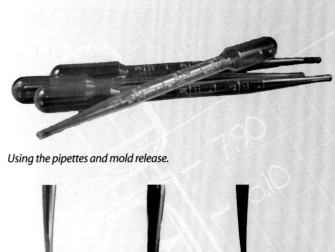

*Using the pipettes and mold release.*

*Prototype and Replica Part.*

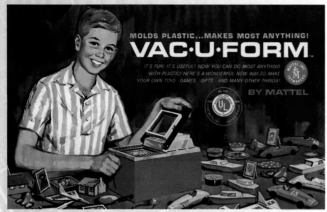

Mattel Vac-U-Form.

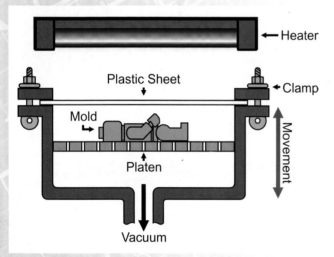

Plastic Forming Diagram.

Examples of Jimmy "THE BOXMAN" Chavez's packages.

Mattel Vac-U-Form™. This was a simple vacuum former that allowed someone to head up a plastic sheet and then swivel it over on top of a mold and vacuum out the air with a simple hand pump forming a new toy.

## Basics

Vacuum forming remains a popular deforming process where the vacuum removes the air underneath a soft and flexible thermoplastic sheet creating pressure to pull the plastic onto a mold. The vacuum forming process starts by raising the plastic sheet towards a heater to soften the plastic before being pulled down onto the mold create a draping form. The aim is always to create a high definition outcome without any excessive thinning of the plastic taking place. Vacuum forming is used to make many toy containers as well as several other commercial products.

## Vacuum Forming Small Parts

Vacuum forming is one of the most economical ways to create custom parts after the purchase or creation of the vacuum former. For small part creation a dental vacuum form machine is the best option due to the type and power of the vacuum motor. It is simple to redesign molds and parts can be easily duplicated.

## MATERIALS

Small dental vacuum form machine
Carving tools
Wood or high density foam (Renshape Foam)
Emery board or sandpaper (recommended 400 to 600 grid)
Small utility knife
5.5 x 5.5 15ml plastic
Heavy duty silicon lubricant spray
Paints and/or water-slide decals

**Step One:** Carve a mold out of the wood or high density foam. I use an emery board or sandpaper to smooth out the edges and perform any fine tuning. It is important to minimize any undercuts which would make it more difficult to remove the mold from the cast.

**Step Two:** Spray the mold with a thin layer of heavy duty silicon lubricant, this allows the mold to be easily released.

**Step Three:** Preheat dental vacuum form machine for three minutes.

**Step Four:** Place 5.5x5.5 15ml plastic into the dental vacuum form machine by securing into clamps.

**Step Five:** Put the mold onto the center of the vacuum surface. It may be necessary to surround your mold with a small barrier to prevent webbing (excess gathering of plastic). This can be easily made from scraps of wood.

**Step Six:** Allow the apex of the plastic to droop. For this example, I let it drop an inch and a half. However, the length is dependent on the height of the mold, the taller the mold the more it should droop.

**Step Seven:** Manually pull down on the handles of the dental vacuum form machine so that the plastic blankets over the mold. Then immediately turn on the vacuum.

**Step Eight:** Allow it to cool for one minute. Then carefully pry the mold loose from the cast.

**Step Nine:** Customize the cast with paints and/or add water-slide decals.

*After the plastic is heated pull it over the mold.*

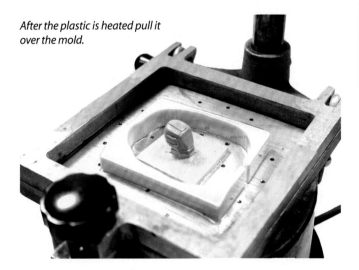

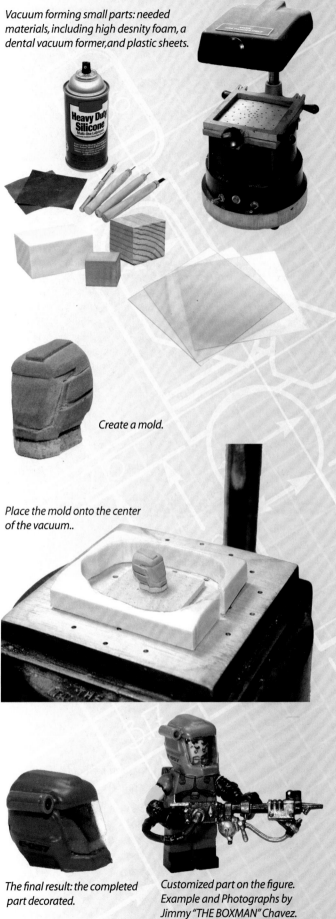

*Vacuum forming small parts: needed materials, including high desnity foam, a dental vacuum former, and plastic sheets.*

*Create a mold.*

*Place the mold onto the center of the vacuum..*

*The final result: the completed part decorated.*

*Customized part on the figure. Example and Photographs by Jimmy "THE BOXMAN" Chavez.*

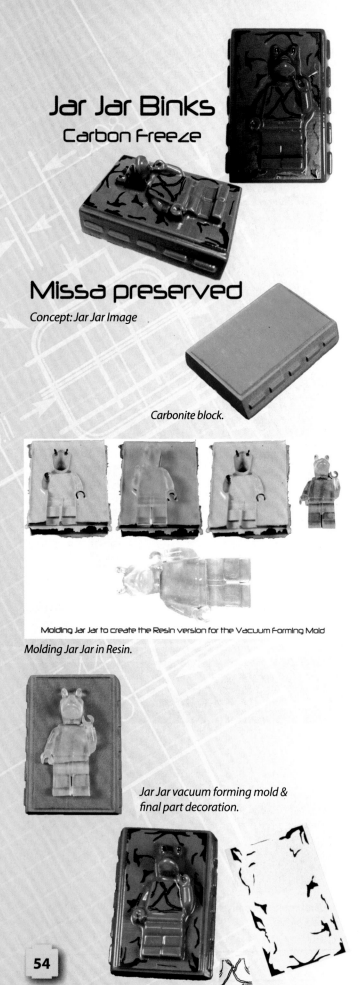

# Jar Jar Binks
## Carbon Freeze

# Missa preserved

*Concept: Jar Jar Image*

*Carbonite block.*

Molding Jar Jar to create the Resin version for the Vacuum Forming Mold

*Molding Jar Jar in Resin.*

*Jar Jar vacuum forming mold & final part decoration.*

# Creating a Vacuum Forming Mold

It is easy to think about all the items that could be made with a former, figuring out how to create the forming molds is a whole new issue. This was one of the greatest difficulties with my learning to vacuum form because I was use to creating parts for silicon molding. Now my mold is 1 sided, thinking this way is a twist. Honestly creating any original part is always the most difficult step and it is commonly the one overlooked as we have our eye's on the finished prized and not the steps we need to conquer to reach that prize. When you take these steps and master them along the way you have a greater sense of accomplishment from the end result and something you can truly call your creation.

Having gathered all the required materials let's roll through the creation process. The goal is to create a vacuumed formed Jar Jar Binks in Carbonite, inspired by the current Han Solo carbonite LEGO part. Who hasn't thought about stashing this *Star Wars* character in carbonite to finally keep him quiet? Simple thought, just take Han off the carbonite and add Jar Jar, however Jar Jar is taller than the typical minifigure. This meant everything had to be scaled up.

Start with the simplest portion of the mold, the carbonite block. By creating the block separately from the figure portion, the figure could be change out to create several variations, ie place multiple people in carbonite freeze. RenShape foam is available in quite a few densities and the most common is similar to bass wood, which is commonly used for carving. I carved out the sides of the carbonite block by hand using an X-acto knife. The hard part was keeping the angle correct and consistent while carving in the carbonite control pads. Then using a 4 inch hobby table saw add the kerf cut around the top edge. The next step was to cut Jar Jar Binks in half, a difficult task to perform safely.

The easiest way to cut Jar Jar in half, mold half of Jar Jar. Created a mold box and used a modified brick 1x1 with a stud on 1 side to suspend Jar Jar off the side of the mold box. Fill the lower portion, thus molding the front half of Jar Jar. After the mold set, poured in some resin and instantly you have cut Jar Jar in half, only needed to clean it up a touch by sanding the back, an activity safe for the fingers.

With the front half of a figure and the carbonite block it was time to vacuum form the final part. Forming was actually the easiest part, because I had taken time and created the forming mold correctly. I had finally gotten the result I was after, now to finish the part, which would take paint, decals, and some scissors. By spraying the interior of the part the part would retain a shiny finish from the outside of the part. The rest is simple, create and apply waterslide decals to replicate the carbonite surface and Jar Jar's clothing. Oh, grab a Sharpie to color Jar Jar's eyes.

So in the quest to vacuum form a Jar Jar in Carbonite the following skill sets were used: sculpting & carving, RTV silicon molding, resin casting, painting, decal design, and decal application. Sometimes the journey is really the most important part. When creating something new it will commonly utilize multiple skills.

For those wanting to build a vacuum former there are plenty of plans online for creating simple formers.

Make Magazine's website is a great resource for these plans:

*http://makezine.com/images/store/Vacuumformer-lo.pdf*

*http://makeprojects.com/Project/Kitchen+Floor+Vacuum+Form er/68/1 http://blog.makezine.com/2012/09/26/dirt-cheap-vacuum-former/*

# 3D Printing

One of the biggest technology advances in minifigure customization, three dimensional printing. Three dimensional printing is a method of rapid prototyping, which involves creating digital 3D objects and 'printing' them by having a printer lay down successive layers of a material. Objects manufactured using this process can often be extremely detailed and include awkward shapes. Other manufacturing processes such as metal injection molding cannot easily manufacture these awkward shapes without multiple molds. The process of 3D printing allows a minifig customizer to physically make a 3D model without having to deal with extremely large production costs and minimum orders. The rate at which these models can be ordered and received is merely a matter of days with companies such as Kraftwurx and Shapeways. This is also a great option for prototyping any mass production piece as all the measurements can be confirmed before a metal mold is tooled.

The materials that can be produced by a 3D printer are extensive. Kraftwurx, the leader in hobby creation of 3D printed parts, currently provides metals, plastics, glass, ceramics and sandstone, actually over 37 different materials. This technology was very limited in the early days, however currently it has limitless potential in the creation of custom items. The production of custom LEGO-like elements first brought me to Shapeways' website. At the time, there were two materials which suited both the detail and texture of LEGO elements; the 'White Strong and Flexible' (WSF) and the 'Detail' plastic. The WSF material is a nylon material and as the name would suggest is strong and flexible. Although cheap, it has a somewhat bumpy surface due to it originating from a powder. The Detail material solved this surface problem. The Detail material is an acrylic based photopolymer, which is a smooth material with a much higher resolution to the printed item. This resolution has limitations as it lacks both the flexibility and the strength of WSF. Over the last year however, many coloured variations of the two materials have been introduced as well as a polished WSF option which significantly improves the problem of its bumpy surface texture. The high demand for such small detailed objects also allowed Shapeways to release the 'Frosted Detail' and 'Frosted Ultra Detail' materials. This UV curable acrylic plastic can achieve details of up to 0.1 mm and is very smooth to the touch. Despite being quite costly compared to the other materials, it is by far the most authentic looking in the hands of a LEGO Minifigure as it most closely resembles LEGO element materials (ABS).

3D objects to be printed are created in 3D CAD package, which can range from being free to costing thousands of dollars. Freeware such as Google SketchUp and Blender can get the job done but for the more experienced modeler, software such as Autodesk Inventor and Solidworks are used. The process of creating a 3D object can vary program to program. The examples below were kindly created by Michael Inglis. A basic tutorial for creating a cube is presented below. Basic shapes can be created in a couple of steps in most any software.

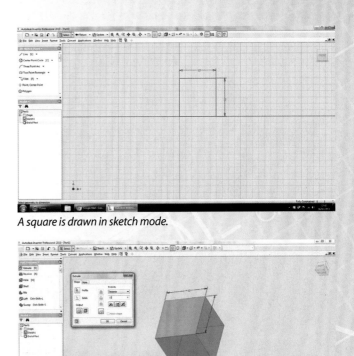

*A square is drawn in sketch mode.*

*The profile of the square is then extruded into the cube.*

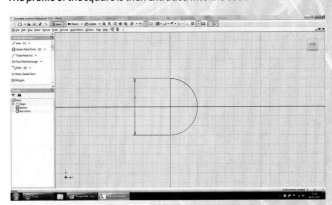

*The process of creating a sphere is somewhat similar:*
*A semi-circle is drawn in sketch mode.*

*The profile is then revolved on an axis, thus creating a sphere. Examples and images by Michael Inglis.*

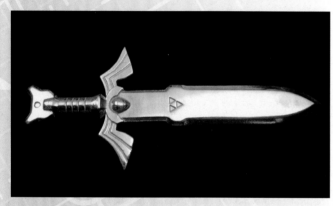

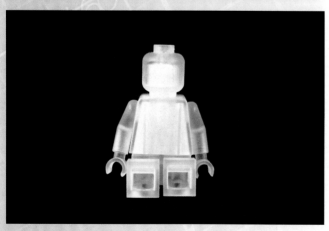

A more complicated tutorial featuring one of Michael's designs can be viewed here – [http://www.flickr.com/photos/62934730@N00/7309053998/in/photolist-c8SP7o-9zGjuA-awRr5G-dPtxW2]

When creating a LEGO scaled model, a series of measurements must be taken from a LEGO Minifigure. These measurements create the foundations of the model to be made. The next step is fitting the design of the model in with the LEGO aesthetic or LEGOverse. Commonly this requires reducing the amount of detail for the item. If you are basing the 3D model on a real life object it will likely require the item to be made more substantial. This is because you are translating the item into LEGOverse space where dimensions dramatically change due to the out of proportioned minifigure scale (Minifigure scale, assume that the minifigure represents a 6 ft tall person, this would make the head roughly 2 feet tall as it is 1/3 of the total height of the figure). Because of the odd dimensions of the LEGO minifigure items will commonly have to be rescaled in all 3 dimensions to make them appear correct in a minfigure's hand. For example, if designing a LEGO minifigure scale flute a reduction in the number of keys and an increase in their size help translate the flute into LEGO space as it would be easier to recognize and create. The scale of the detail also has to be in keeping with the printing companies' minimum wall thicknesses. Minimum wall thickness is the minimum thickness the printer can accurately create an item and support the structure. This is usually around 1 mm in plastics, but if the model is to also be printed in stainless steel, for example, a much larger minimum wall thickness of 3 mm would be required. So when creating your models tolerances are the final specification to be taken into consideration. Although the plastics have a relatively low tolerance, the metals have a much larger tolerance as they often go under some shrinkage in the curing oven. Shrinkage is critically important when creating items such as helmets, make sure the 3D model you create have tolerance for shrinkage. This only really needs to be taken into consideration for helmets as the tolerance on an item's grip area doesn't fluctuate.

Although perfect for testing models before they go onto mass manufacture, 3D printing for LEGO still has a way to go due to the limitation of printing in full color materials. Other than the 'Full Color Sandstone' material and the available dyed WSF materials, it is not possible to properly color the model unless one is to paint or vinyl dye the object (great write up on MAKE on RIT dyeing 3D printed objects). Keep in mind that if the item is placed in and out of a minifigure's hand the paint could easily be rubbed off. The texture of the finished product has come a long way in recent years. The 'Frosted Ultra Detail' material is, as mentioned above, extremely smooth and has only barely visible print lines. These can be easily removed with a sandpaper or a file.

## RESOURCES

### Tutorials
*http://www.softwaretrainingtutorials.com/inventor-2012.php#*
*http://www.maxbasics.com/*
*http://www.blender.org/education-help/tutorials/*
*http://sketchup.google.com/training/videos.html*

### Examples
*http://www.shapeways.com/shops/hobos*
*http://www.shapeways.com/shops/spiderpudel* http://www.shapeways.com/shops/battlefieldbricks_jon

Michael also has a lot of 3D printed examples in his Flickr account:
*http://www.flickr.com/photos/60858662@N07/sets/72157627029210361/*

## 3D Programs

Autodesk Inventor (Educational version free with a student email)
Autodesk 3ds Max (Educational version free with a student email)
Blender (Free)
SketchUp (Free)
Cinema 4D (Cost)
Solidworks (Cost)

# Grey Market Accessories

This book has been primarily concerned with teaching how to create custom figures and all the accessories needed to outfit your figure. This isn't always practical and until recently (with the creation of the Collectible Minifigure Series) LEGO's accessories were fairly limited. So when you can't or don't want to create every part of the figure from scratch, it is time to look to the grey market of LEGO compatible accessories. This is actually quite a large field with about **6** main contributors beyond the LEGO clones: KRE-O, Bestlock, Cobi, Megablock, Character Building, and Oxford.

To begin this discussion please consider that many of the accessories created by the clone Brick and Action figure companies including: Bestlock, Character Building, Cobi, Games Workshop, Hasbro, Medicom Kubrick, Megablock, KRE-O, Oxford, Sidan, Stikfas, and many others are very compatible with LEGO figures. Many customizers have used hats, weapons, capes, and other odd parts to complete a custom figure. One of the easiest ways to make a LEGOized *Star Wars* Bouush Leia figure, for example, is to use the Hasbro Action Figure Helmet over a LEGO head (This was the method of choice before LEGO finally made this figure). This can be a very economical way to get the needed accessory, especially if you don't have the time to make it. Buy the action figure or yes one of the other companies' sets. I know many wouldn't touch their inferior bricks, but sometimes it is the only way to get that needed accessory item (Also you can use those bricks to build your molding boxes with instead of destroying your LEGO bricks, so you can repurpose the bricks). Commonly many of these "other" companies' accessories are sold on eBay and a few, like Sidan, are readily available and have created specific websites for their accessories direct sale (http://www.minifigcat.com/). Several of these other companies have themes in line with LEGO and where they succeed, in my opinion, is more artistic accessories. Instead of a plain straight spear it might have detailing or be slight crooked. This detail could help make your custom figure unique.

With knowledge of the LEGO Clones and Action figure market we turn to the Grey Market, manufacturers that specifically create accessories that are compatible with LEGO figures. To summarize I have created a table below of all the vendors, their specialty, and their store location. This is not an exhaustive list of where you can buy items; many of these groups have distributors. To locate a distributor close to you please check their websites or run a Google search for their names and you can find one in closer proximity to save on taxes and possible import fees. However, whenever possible I always try and buy directly from the manufacturer for the best service. Navigating these sites can be very time consuming, understanding your needs before visiting can be helpful and I suggest using search functions on the sites when possible.

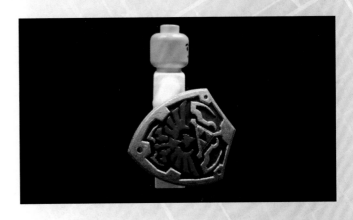

# Third-Party LEGO Compatible Manufacturers

| Manufacturer | Owners | Web Address | Decals | Parts | Cloth | Mods |
|---|---|---|---|---|---|---|
| Amazing Armory | Hazel | bricklink. com/store. asp?p=AMAZING ARMORY | | X | | |
| Arealight | Bluce Hsu | Arealightcustoms. com | | X | | |
| BrickArms | Will Chapman | brickarms. com | | X | | |
| Brick Brigade | John Canepa | brickbrigade. com | | X | | |
| Brick Display Case | Kwok Chi Keung | brickdisplaycase. com | | X | | X |
| BrickForge | Kyle Peterson | brickforge. com | | X | | |
| Brick Fortress | | brickfortress. com | | X | | |
| BrickStix | Grayson MacLean | brickstix. com | | X | | |
| Brick Warriors | Ryan Hauge | brickwarriors. com | | X | | |
| Brickmodder/Lifelites | Rob Hendrix | brickmodder. net, www. lifelites. com | | | | X |
| BrickTW | Kevin Chu | shop. bricktw. com | | X | | |
| Cape Madness | | bricklink.com/store.asp?p=dingham | | | X | |
| Citizen Brick | | citizenbrick. com | | X | | |
| Combat Brick | | combatbrick. com | | X | | |
| Custom Brick & Minifig | Christo | myworld. ebay. com/christo7108 | | X | X | |
| Fine Clonier | Jared Burks | fineclonier. com | X | X | | |
| Little Armory* | Jeff Byrd | littlearmory. com | | X | | |
| little-legends.com | Darin Braund | little-legends. com | | X | | |
| MINGLES | Michael Inglis | shapeways.com/designer/MINGLES | | X | | |
| MinifigCat | Sidan | minifigcat. com | | X | | |
| MinifigCustomsIn3d | Andreas Holzer | shapeways. com/shops/MinifigCustomsIn3d | | X | | |
| Minifigmaker | Rick Andrews | minifigmaker. com | | X | | |
| MMCB | Mark Parker | mmcbcapes. servaus. net | | | X | |
| Roaglaan Stickers | Tim Fortney | roaglaanscustoms. com | X | | | |
| Saber Scorpion | Justin Tibbins | saber-scorpion. com | X | | | |
| The Clone Factory | James Anderson | flickr.com/people/theclonefactory | | X | | |
| The Little Arms Shop | | thelittlearmsshop. com | | X | | |
| Unknown Artist Studio | Victor Sobolev | unknown-artist. com/product | X | X | | |
| V&A Steamworks | Guy Himber | www.crazybricks.com | | X | | |

*Please note: Descriptions of the various third-party vendors and their wares were provided by said vendors. This article and the author of this book does not endorse or vouch for any of these manufacturers. If you have concerns, pleas contact the site of interest's customer support people before purchasing. Also note the author owns the Fine Clonier.*

## Amazing Armory
Amazing Armory offers a line of highly detailed military pieces that range from helmet accessories to highly detailed weapons. Most with a Science fiction/video game twist.

## Arealight Customs
Arealight Customs make fun and high quality custom accessory for anyone who enjoys customizing their own minifigure creation. Custom parts are made of high quality ABS and many are available pre-printed.

## BrickArms
BrickArms sells WWII, modern, and Sci-Fi weapons, accessories, and helmets. All are made of injection molded ABS, and are available in multiple colors including gunmetal and electroplated chrome. BrickArms also sells custom-printed minifigures, complete with matching accessories. BrickArms sells over 100 different weapons & accessories, weapons packs, and custom minifigures.

## Brick Brigade
Brick Brigade is a site dedicated to the creation of quality Custom LEGO Military Models and Soldier Minifigures. Taking LEGO parts, and creating unique vehicles, planes, weapons, vessels, structures, equipment, and soldier characters complete with custom printed decals, weapons, gear, and easy to read instructions. We draw inspiration from the heroism of military fighting men and women, which have inspired hundreds of stories in classic books, and films.

## Brick Display Case

Brick Display Case is a site offering an alternative to the limited LEGO display case. Brick Display Case is specially designed for minifigures display. It works for various brands such as LEGO, Mega Bloks, and Kre-O.

## BrickForge

BrickForge provides the community with a unique assortment of custom minifigure accessories. Their catalog of over 100 specialized elements and 20 vibrant colors spans many themes including: fantasy, historical, modern and sci-fi; allowing minifigure enthusiasts to construct a variety of impressive characters.

## Brick Fortress

Brick Fortress offers short (Stubbie) legs that articulate just like regular sized LEGO legs.

## BrickStix

Dreamt up by a kid, the award winning BrickStix® clings and Mod™ stickers are designed for brick customization and reusability. Brick-Stix are reusable, removable and restickable cling decals. They work using static cling. Mod are removable and repositionable stickers. Both products enhance LEGO® creations and can be easily removed without damaging the brick. "We've got your bricks covered!"

## BrickWarriors

BrickWarriors offers a wide range of minifigure-compatible weapons, helmets, and armor. Our products range across all themes, all the way from Greek Mythology to 1920s Gangsters to Sci-Fi. "All of our pieces are designed with the highest tolerances and are injection molded with ABS plastic, the same plastic that LEGO® uses, in order to ensure that they fill fit seamlessly with your minifigures."

## Brickmodder/Lifelites

Lifelites Custom LED lighting kits and accessories allow for lighting modifications to be added to your MOCs, minifigures, and other scale models.

## BrickTW

BrickTW come from Taiwan who are partners with a deep emotion for LEGO. They hope to create a paradise and invite all LEGO fans to join. Their focus is in the Asian historical theme, but they have a lot of innovative items too. They offer 480 components in different colors; resulting in an amount around 288,000 items in their store.

## Cape Madness

The Ultimate in custom made capes and cloth accessories from CapeMadness are now available online thru bricklink, not just at Brick Conventions.

## Citizen Brick

*Citizen Brick* is your online resource for impeccable custom-designed small toy collectibles. We know there are a lot of places to shop for these types of curios, and we also know that you appreciate quality craftsmanship.

## Combat Brick

The online store dedicated to custom LEGO® Military creations. Here you will find a variety of Army builder model kits, custom LEGO® minifigs, toy weapons and guns.

## Custom Brick and Minifig - Christo7108

Custom Brick and Minifig creates a very high quality of custom LEGO scale items. These items include custom capes, glow in the dark blades, head accessories and excellent custom figures. All our items are printed and are of a very high standard. Christo7108 is always ready to do new and exciting designs. Complete figures are available, but as they are only sold through eBay the price can vary from auction to auction, so be patient if you are after one of their parts or figures.

### Fine Clonier

The Fine Clonier, my site, offers high quality waterslide decals for several genre as well as custom cast and ABS elements. We strive to incorporate a style similar to LEGO when appropriate and deviate where we deem necessary. We offer thousands of designs which can be combined in infinite number of ways with LEGO elements to create an endless supply of custom figures.

### Little Armory

The Little Armory, which I believe is now closed, was the first ABS parts supplier on the grey market that I am aware of. They created simple yet elegant versions of the *Star Wars* weapons to outfit your custom figures. Many of these accessories are still highly sought after today.

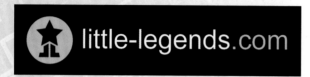

### Little-Legends

We offer a wide range of customized figures, with themes including Modern Military, SWAT Teams, WW2 Soldiers, Gamer Style figures, Zombies, Music and Film. We also stock a good selection of custom guns, army and SWAT armor and headgear, plus various musical instruments.

### Mingles

Custom LEGO compatible accessories inspired from a variety of video games and films. Ranging from an Assassin's Tomahawk, to a Judge's Helmet, with a huge pile of swords in between!

### MinifigCat

SI-DAN TOYS design original of many high quality with special functions custom helmets, armor, belts, weapons such as helmets with night vision goggles, armor with pouches and clips, belts with holsters, rotatable Minigun, etc. Our catalog of over 100 special parts and 20 colors spans many themes including SWAT A-Team, NAVY SEAL, Sniper, Sci-fi, Samurai and Ninja. We use some different materials to design custom parts such as leg panel, goggles cover and gun slings. We offer all you need on the battlefield.

### MinifigCustomsIn3d

MinifigCustomsIn3d provides custom accessories in minifigure scale. The shop contains mainly hats for different topics (Napoleonic, Middle Age, Ancient, Fantasy, Military) and a small number of additional items like rifle, pistol, drums, mace, and more. Currently the shop has nearly 150 different items.

### Minifigmaker

At Minifigmaker.com, collectors and fans can purchase hand crafted, custom designed minifigures. Specializing in customized *Star Wars* and Gaming minifigures, every figure is made with attention to detail and durability for the collector. A variety of materials and techniques are used to make each and every individual figure. This may include custom designed water-slide decals, paints and sealers, fabrics, custom molded helmets and other specialty parts.

### MMCB

MMCB Custom Minifig Cloth Accessories specializes in making fabric accessories for standard LEGO Mini-figures. Their designs range from basic capes through to highly detailed pauldrons, holsters, and tents. Each piece is made from specially treated fabric closely matched to LEGO colors. They currently have 90 different designs which are available in 33 colors. Most of these can have additional detailing (borders, emblems, camouflage) to make a total of 688 individual products.

## Roaglaan Stickers

Roaglaan Stickers was purchased by Brickmania, a company specializing in custom military vehicle sets. Most of Roaglaan designs are based on World War II and modern military themes. Designs are provided as stickers or water-slide decals.

## Saber Scorpion

Saber Scorpion's lair offers sticker sheets to create your favorite custom figure across many different films and games.

## The Clone Factory

The Clone Factory creates custom molded parts and complete figures, primarily in the *Star Wars* genre.

## The Little Arms Shop

There is a cloud of mystery around how this shop started. The owner use to be a distributor for the Little Armory and then suddenly cut the Little Armory out of the picture. They offer many of the same items that the Little Armory did, so if you are desperately looking for one of their items check here.

## Unknown Artist Studio

Unknown-Artist-Studio is hobbyist owned and run online shop catering to AFOL's and customizers around the world. They offer intricately fitting fabric cloaks, coats, and holsters and sheaths that work with official and aftermarket elements. In addition to fabric, they also have resin cast accessories. Requests for custom colors and elements are always available.

## V&A Steamworks

Custom parts. Known for Stove Pipe Hats and Crazy Arms. These items are machined individually from Aerospace plastics and are perfect for enhancing your Brick Scale Figures and MOCs. Find me on flickr.com - user - V&A Steamworks.

It is readily apparent: the Grey Market is a large and growing with many people making niche items. Understand your needs before shopping and buy only what is needed or you could go broke quickly as many very interesting items are being produced.

*PLEASE NOTE: This article and the author does not endorse or vouch for any of these manufacturers so if you have concerns please contact their customer support people before purchasing. Also note the author owns the Fine Clonier.*

*LEGO Minifigure Stand.*          *LEGO Sports Plate.*

*Display Board.*

LEGO Display boxes.

After creating a custom figure there are still two critical aspects to the process, displaying that custom figure and sharing images of that figure online. This chapter will cover displaying, the next will cover photographing and sharing. Even if you don't have a large selection of custom minifigures perhaps you wish to display your official figures. How can this be done, well there are multiple products out there and multiple ways to use them. Minifigure Customization: Populate Your World discussed many of the available options. MC² will show some of the newer options, but will focus on customizing the options to give them an added pop to showcase your custom figure.

Before we can discuss sound, lighting, decaling, or other customized options we need to look at the available display options. With all things, let's start simple and work our way to the complex. Let's review of the available display options by looking at what LEGO has given us to use to display our figures. I bet there are more than you realize and these options are typically the ones that best integrate into our LEGO collections. The most recent display stand that LEGO has given us accompanies the new "Minifigures" line. This simple little plate (Tile, Modified 4 x 3 with 4 Studs in Center: *www.bricklink.com/catalogItem.asp?P=88646*), which is quite effective for displaying figures and is economically priced around twenty cents on Bricklink.

This plate is also similar to the display plates that LEGO gave us a few years ago. These tiles (Modified 6 x 6 x 2/3 with 4 Studs and Embossed Letters: *www.bricklink.com/catalogList.asp?pg=3&ca tString=38&catType=P&v=2*), came with a few sets and had the embossed words "Star Wars,""Sports,""Rock Raiders,""Ninja," or "City." These came in various colors and all but the "Star Wars" versions can be found for pennies on Bricklink making it an economical way to display your figures. This little stand features a card slot which means you could display your figure with a nice little printed backdrop or figure schematic. Using these and a special board I had made makes for a great display. The plates slide into a special slot cut into the board that makes display shelves for the figures, seen at left.

Schilling created (in 2011) a LEGO display case that is great. It is sealed to avoid the dust issue and hangs on the wall nicely. It also has feet and stacks. There are a variety of colors and two sizes. These will be great for individual display or groups of figures. These can be expensive and hard to find. I get mine from Toys R Us when they have a sale. However, let us not forget that LEGO has also supplied us with magnetic stands (Magnet, Brick 2 x 4 Sealed Base with Extension Plate with Hole: *www.bricklink.com/catalogItem.asp?G=74188*). These are likely the most economical currently as they are in high supply on Bricklink. They are also available in several colors. Much like my display boards described above these could be used in several creative ways to display figures. Just think of what could be done with these and a magnetic dry erase board. To take it one step further, there are now magnetic paint primers. You could paint the wall of your LEGO space with a magnetic primer, use a colored paint over the primer and then display all your figures without placing a single nail in the wall by using these magnetic stands, this is how I have my LEGO room set up.

These are the simple stands that LEGO has supplied. Using bricks and tiles you can create individual displays stands, risers, and even display boxes. The best LEGO brick built display for a figure I can think of is a vignette. These are small scenes that capture the nature of the figures housed inside.

There are many commercial options available to use to display your figures. There are small plates available from Minifig World (*www.minifigworld.com/*), as well as a few variations available from some other manufactures on the secondary market. If you want to keep your figures dust free there is a recent addition is from Brick Display Case (*http://brickdisplaycase.com/*). The case from Brick Display Case features a wall hanger and is also stackable. It is more compact that the Schylling LEGO product, but holds many figures and allows for more action posing of figures, more on these cases shortly. There are also small acrylic boxes, and domes available to complete protect your figures. These can be combined with the brick built LEGO stands to house your figure.

There are a few larger acrylic boxes that are quite useful and widely available from the Container Store (*www.containerstore.com/shop/collections/display/cubesCases*). One that I particularly like has internal dimensions of 6 by 36 studs, which is no longer availableL. This makes it perfect to display figures, especially if using the embossed LEGO plates mentioned above. This is what I use to house my autograph collection. As this is not longer available I have found a new product on the market from BrickDisplayCase.com. It will hold 8 figures right out of the package, however it allows for the addition of LEGO plates, which then gives it significantly more display space (6 x 41 studs). These cases stack, have wall hangers, and can easily have light effects added, see the next section. This case can even be built into structures due to it's studs on top and acceptors on the bottom.

As you can see there are many ways to display and protect your figures. Get creative and look at some of the options above and try to think outside of the box. The best displays are the ones that are a touch creative and incorporate the display case into the figure.

LEGO Magnetic Display & Magnetic Paint

Zoe the Zookeeper, mascot of the Houston Zoo.

Brick Stand, photo by Jared Burks

Display Stand, photo by Mark Parker

Autograph display.

Acrylic Boxes:Photos by Matt Sailors (theinformaticist)
The clear boxes with the black bases are from AMAC Plastics (www.amacplastics.com). These are in the Showcase Series from AMAC Plastics. AMAC Plastics sells wholesale with a $100 minimum order, however the cases are available at some retail outlets online. The figure boxes shown are models 802C (the 3x3 insert size) and 805C (the 4x4 insert size).

Brickdisplaycase.com photo.

Epoxy lens/case
Wire bond
Reflective cavity
Semiconductor die

Anvil
Post } Leadframe

Flat spot

Anode          Cathode

*LED diagram. Note directionality of power flow*

POLICE PUBLIC CALL BOX

*LEDs are small enough to place in an R2 figures dome, or at the top of a TARDIS.*

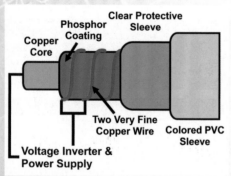

Clear Protective Sleeve
Phosphor Coating
Copper Core
Two Very Fine Copper Wire
Colored PVC Sleeve
Voltage Inverter & Power Supply

*Diagram of EL wire. Note wiring of two small copper wires to one battery pole with the other pole connected to the copper core.*

# Creating Display Figures and Customizing Display Cases with Light

When adding lights to a minifigure or a minifigure accessory the playability is significantly reduced because the figure now has wires and possibly a battery running through it to power the light. In my opinion this creates a display figure. As such these topics will be discussed in this section. Once the basics on creating a display figure are shown the topics will be expanded to show how to light a display case, add sound, and other effects to further customize the display of your figures.

This is really an area of customizing all its own; it has the ability to completely transfer an awesome figure, into a completely unforgettable figure. LEGO has even tried to accomplish this area of customization, however not very successfully. This area of customizing can be a bit overwhelming as it involves electricity and batteries, which mean caution, must be used. Hopefully after reading this section you will feel more comfortable attempting a lighting project on your favorite figure. Lighting a figure is typically done by one of two sources, LED or EL wire. This is a very complicated set of modifications; as such this section will cover the basics and give one example with some photos of what can be done. Later an additional, more advanced article will cover a more involved project. Please note, unless kits are used soldering is required for many of these projects. This is an advanced skill and beyond the scope of this article. There are many great tutorials online (one is noted in reference section), please review a tutorial before attempting yourself.

LEDs, or **L**ight **E**mitting **D**iodes, have been around since the early 60s and have recently become very small and very inexpensive. You will find them everywhere from flashlights to lightbulbs to TVs and even in stores like Ikea. They come in a wide array of colors and sizes. These little lights practically never burn out and use very little power, which make them perfect for LEGO lighting projects. Most often LEGO projects use small surface mount (SMT) LEDs (common size for minifigure projects are 1.25 mm x 2 mm). These can be found very economically on eBay in small quantity. Single LEDs can easily be powered by a small watch battery (size CR1025). Because LEDs can be powered by such a small battery, these can be contained inside the minifigure body. With a rotary tool the torso can be converted into a battery pack. This can be done with fuse taps and a small switch. Part of the interior of the minifigure torso and hips needs to be removed to make room for these parts. All the details about this sort of lighting are in a tutorial on the Brickmodder website, so while basic details are listed here, I suggest visiting Rob's website for a very helpful step by step tutorial link in the resource section.

EL wire is a much newer product and really came to the forefront through its use in the film TRON Legacy. EL wire, or **E**lectro**L**uminescent wire, is a thin copper wire coated in a phosphor which glows when an alternating current is applied to the wire. This sounds quite complicated, but there are small voltage inverters that alternate the current and merely plug into the length of the wire or are incorporated into the battery holders. The really cool feature is that because the current alternates direction, the EL wire can dead end and not be connected at both ends of its length. Look at figure 3; see the end of Wonder Woman's lasso. EL wire gains its color from the PVC sleeve applied to the wires, thus EL wire merely emits a white light unless coated. EL wire also comes in various thicknesses, with the thinnest being approximately 0.9mm in diameter.

So now that the light sources have been introduced we need to understand the main differences. A LED is a simple lighting device that merely needs a switch, battery, and battery connectors. However, EL wire needs a switch, battery, and a voltage inverter. The biggest difference is the voltage inverter, which is critical and needs some discussion before this article proceeds. A voltage inverter allows the EL wire to illuminate. It raises the voltage, while lowering the amperage and creates an alternating current. Because of the necessity of this device EL wire requires an external battery pack as the inverter is too large to be contained inside the figure body. However, with some planning a small base can be constructed out of LEGO to conceal the battery/ inverter pack. The other key to EL wire is that the length of the wire dictates the amount of power required to properly light the wire. For most minifigure projects 1-1.5 feet is plenty, which can be easily powered with 1-2 AA batteries (1ft or less can be powered by two small watch battery AG13 size). If you have larger projects using EL wire, please refer to the resource section below for additional information.

*EL wire lasso for Wonder Woman.*

So now that the basics have been covered the remainder of the section is going to show how I recently converted Jack "Ewok in Disguise" Marquez's TRON Lightcycle with EL wire. This is a nice simple start to using EL wire. Jack was very nice to supply instructions for his design on Flickr, which I modified slightly for my version. His instructions can be found on the following page.

Now that you have seen that LEDs and EL wire are simple to use and that many kits are available, many custom figures, vehicles, and lighted vignettes can easily be created. Please show us what you can create.

## RESOURCES

### Tutorials
Rob "Brickmodder" Hendrix:
Lightsaber Modification Tutorial: *http://www.brickmodder.net/ tutorials/howto/BFDC04%20-%20Brick%20Modification.pdf*
EL Wire Tutorial: *http://www.ladyada.net/learn/el-wire/*

Jack "Ewok in Disguise" Marquez Tron lightcycle instructions: *http://www.flickr.com/photos/28177764@N07/ sets/72157626348735081/with/5609117928/*

Basic Soldering: *http://www.aaroncake.net/electronics/solder.htm*

### Examples
Light-up Minifigure Flickr Pool:
*http://www.flickr.com/groups/1274076@N23/*

LifeLites Flickr Pool:
*http://www.flickr.com/groups/1221775@N25/*

### Supplies
**LEDs:**
*http://www.ebay.com* Search for LED and size needed
**Fuse taps:**
*http://www.radioshack.com/product/index.jsp?productId=2102780*
**Switch for LED:**
*http://search.digikey.com/us/en/products/AYZ0102AGRLC/401- 2012-1-ND/1640121*
**LED kits:**
*http://www.lifelites.com/*
**EL wire, kits & inverters:**
*http://Thatscoolwire.com*

*EL wire TRON Light Cycle, Illuminated with 1.5 ft of EL wire and one AA battery box/inverter.*

# Step by Step Lighting of the Lightcycle

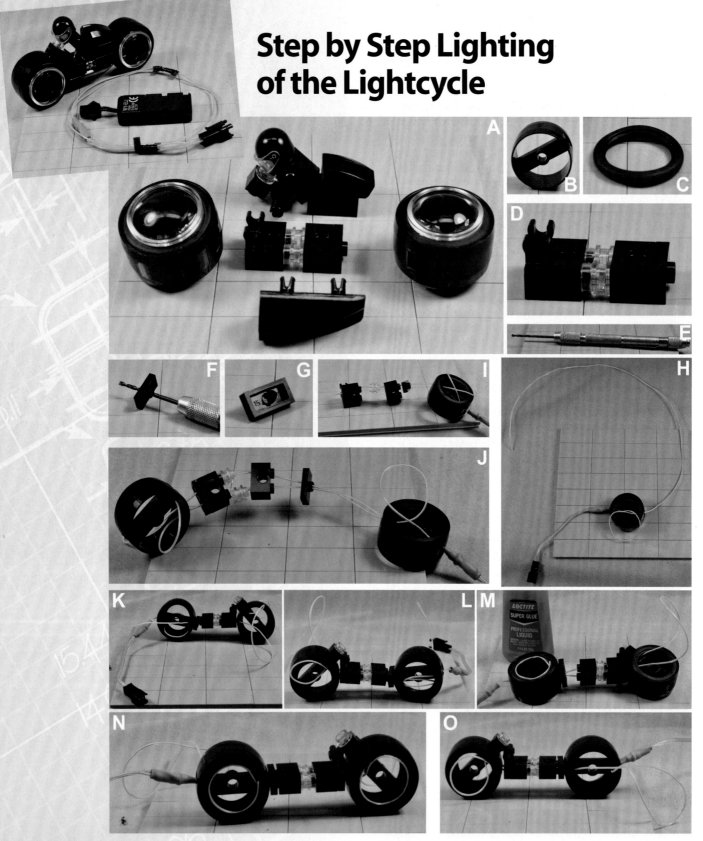

A. Exploded view of light cycle, note center core of cycle. B. Wheel. C. Inner portion of wheel, LEGO bicycle tire, note groove, this is critical for holding EL wire. D. Center core needs to be drilled out. F. Jumper Brick is drilled through center. G. Bottom of Jumper Brick has small "poll" removed. H. Enter hole in back of wheel, pass through and loop. I. Traverse down one side of center core. J. Drill two small holes in front wheel. Look and travel back down other side of center core into back wheel. K. Loosely assemble center core. L. Starting from the dead end tip of EL wire start loop. M. Superglue tip into grove in back wheel, spot glue along loop. N. Glue in front wheel. O. Glue in other side of front wheel. Pass excess EL wire out back, glue in back wheel. Reassemble Lightcycle.

## Lighting Minifigures and Displays

Now that we have discussed the basics on how to light a figure or a figure accessory, how do you light a display? The biggest thing to remember is the accumulated voltage and amperage required for the LEDs you choose to use. This all depends if the LEDs are wired in series or parallel. Yes, individual LEDs can get complicated if you need to use multiple to achieve your lighting affect. This can be alleviated by using either EL wire or LED strips.

EL wire has been covered above; the trick for longer lengths is simple that you need a larger inverter. Most sites that offer EL wire also offer the inverters required to power the EL wire. Simply make sure you are purchasing an inverter that matches or is slightly larger than the length of LE wire you are purchasing.

If purchasing the LED strips make sure you are purchasing what you need, check the color temperature of the LED. The temperature affects the color of the LED, be sure to check the details when purchasing. For LED strips, the best prices I have seen are on eBay. Since these are going into a display case in this example, there is no reason to get a waterproof version, but these do exist. There are also RGB LED strips that would allow you to alter the color to most any color. These also come with a controller. Remember the amount of space you have in the case.

One SMD (surface mount diode) LED was added to each compartment. Each is wired directly back (parallel) to the 12 volt battery box (8-AA batteries - 8x1.5 = 12 volts).

The case was cut between compartments to allow the light strip to pass between each compartment. Since this is already wired, you simply wire the positive strip one to positive strip two, same for negative, then back to battery, simple. These strips come in voltages, so match your power supply to your strip power requirements. As I mentioned there are several control boards for these strips.

Check out these videos of what is possible:

RGB Strips controlled optional controller using the Brickdisplaycase. com case. This video walks through all the steps of wiring and adding the LED strips to a case, these are the same steps regardless of the cases. Thanks for the video goes out to Kwok Chi Keung: *https://www.youtube.com/watch?feature=player_ embedded&v=gxr3nCeKTrM*

RGB Strips controlled by Arduino: *http://www.youtube.com/ watch?v=3QKmsMrsM3g*

## Adding Sound

Adding sound can be done quite easily. There is a huge market of sound chips for cards and toys out there. These are simple devices that allow you to push one button to record and another to play the recorded sound. There are others that connect via USB to transfer MP3s or other sound files to the chip. These typically feature small speakers. The sound chip can then be built into the MOC and with a simple button push the sound can be heard. No wiring required. These are easily found on eBay. Be sure to consider the length of time for your sound file, the size of the board you can embed into your display or MOC.

*Example LED Strips. The left Is an RGB LED strip. The right strip hows the details. Notice that the strip has marked cut points for easy shorting and wiring. Also notice that each section contains the proper resistor to protect the LED from being burned out by adding too much voltage..*

*Parallel wired 12 volt LEDs added to a display case..*

*LED Light strip added to a display case. Photo and case by Rob Hendrix.*

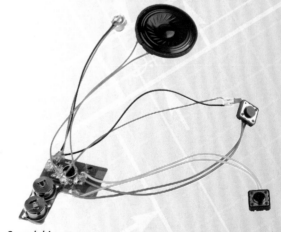

*Sound chip.*

*Photography Setup: Using light sources on opposite sides reduce the presence of shadows. Photograph by Michael "Xero" Marzilli.*

Photographing and sharing your custom figure and all the hard work you did to create it is critical. Don't have a fancy DSLR or even a digital point and shoot camera?? No problem! Almost everyone has a cell phone, tablet, or hand-held gaming device that has a camera. Here are a few suggestions on how to get the best pics possible out of whatever device you have. Remember, the best camera is the one you have with you! Several different tips and tricks will be presented to help capture your work the best way possible.

## Take Care Of Your Camera Phone/Device

Keep the devices' camera lens clean, especially free of fingerprints. Most of us carry our cell phones in our pockets with keys, change and several other items. Dirt, pocket lint or any number of other items can scratch, smear on or cover the camera lens. The first step in a clear image is shooting through a clean lens!

## Lighting, Lighting, Lighting!

To take good pictures with your phone/device you need light, and a lot of it. The reason is that the sensors inside phones/ devices are very very small compared to the ones you find in point & shoot cameras and big DSLR's. Smaller sensors capture less light which leaves the potential of getting a lot of "noise" or grainy unclear images. The more light you have available the clearer and brighter your pictures will be. However avoid getting direct light into the lens or putting your subject too close to a light source. Your pictures will be over exposed, meaning one side is too bright while the other is too dark. With that being said, If possible, avoid direct sunlight.

## The Closer to Your Subject, the Better

One of the most common mistakes with images is that their subject ends up being a tiny, unrecognizable object in the distance. Remember that most LEGO figs are approximately 2 inches tall to begin with. It's difficult to see and appreciate the detail in a figure if we can't see it! Phone/device images tend to be small due to low resolution of the camera (although this is changing) – so fill up your view finder with your subject to save having to crop in on the subject in editing it later (which decreases quality even more). Also, do not use your digital zoom function. This will decrease the resolution of the original photo and there's no amount of editing that will correct pixelated low resolution.

Having said this, getting too close on some model camera phones creates distortion and focusing issues (particularly if the phone/device doesn't have a macro setting). To avoid blurry or tiny pics, find that distance 'sweet spot' to ensure the best quality. This will differ from device to device so practice is key!

## Stabilize Your Phone/Device.

While you're taking a photo, put the device or your arm on a flat, stable surface to make sure the photo doesn't come out blurry. If you can't find a stable surface, hold your arm against your body. A twitch, a cough or even heavy breathing could be the difference between a sharp and blurry picture. Keep Track of the Shutter Release. By shutter release, I mean the time it takes for the camera to capture the image after you press the shutter button. If the shutter button is on a touch screen, the shutter will probably get tripped after you lift your finger. More often than not, such 'tripping' can be cured by keeping your hands as steady as possible. Many devices have a secondary shutter button, so play with all the options and use the one that gives the clearest photo.

## Adjust the Settings

When it comes to photo settings, most phones & devices are now imitating the same features as that of high-end cameras. Of course, the effectiveness and desired effect varies. The most basic setting that should be utilized if available is the Macro setting (previously mentioned). Typically this is indicated by a tiny flower icon. Make sure to select this option to alter the camera's focal point to a shorter distance.

The next setting to be concerned about is ISO, which is an older term that denoted the speed of the film in relation to its sensitivity to light. While digital photography doesn't use film, a similar value is still used. The ISO setting, by default, will be set to "auto." When shooting in direct lighting, you can set the ISO to the lowest value (to avoid grainy, or "noisy", images). But in poor light, use a higher ISO. Keep in mind the higher the ISO number the longer the shutter stays open, which in turn means you must be VERY still while taking these low-lit pictures; Another reason that lighting is so important.

Another critical setting is HDR (High Dynamic Range) imaging. This method captures a greater dynamic range between the brightest and dimmest areas in a field of interest or subject. Using this setting gives a more accurate range of the intensity levels found in the figure and will help when it comes to editing. Non-HDR cameras take pictures at one exposure level with a limited contrast range, which results in the loss of detail in bright or dark areas of a picture. HDR compensates for this loss of detail by taking multiple pictures at different exposure levels and intelligently stitching them together to produce a picture that is representative in both dark and bright areas.

*Changing the bulb color in your lamp is an easy way to get very different affects in your photographs. Photograph by Michael "Xero" Marzilli.*

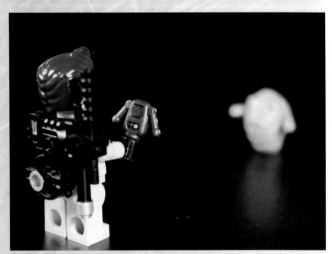

*By controlling the depth of field visual interest is controlled. Use this to highlight those little details that make your custom figure unique. Photography by Michael "Xero" Marzilli.*

## Apps are Awesome!

It doesn't matter if you have an iPhone, Android, or Windows phone there will be hundreds of apps available for your phone/tablet which will enable you to really take advantage of your camera. Some can completely replace the camera app on your phone and offer some amazing features and tricks which you didn't think could even exist. Download a number of free or pay photo editing apps and start editing your pics till your heart's content. You may not be an expert in Photoshop but that doesn't mean you can't let an app do the work for you. Editing apps let you adjust a number of settings and add filters to your pics, providing dramatic effects. Try them out and find the one that works best for you.

## But If Apps Don't do The Trick..

While it's fun and easy to use your phone or devices' built-in or downloaded editing software, editing pictures on your computer will almost always produce better quality images. Remember to always take your shots in color and at the highest resolution possible to keep all your editing options open.

## Computer Photo Editing: The Basics

Now you have captured a few images of your favorite new custom figure and you really want to make sure the pictures are the best you can share. Load them onto your computer from your camera and start editing. Editing is where a lot of experimentation can occur; to convert your color image to black and white, to darken the image, to crop out excessive background, it is all possible and only limited by your imagination. There are many programs and Apps (mentioned above) on the market for photo editing, with the leader in the field being Adobe PhotoShop.

Just like the Apps, there are several freeware/shareware programs that can be used as well (Check www.tucows.com). One that I particularly like is called Irfanview (www.irfanview.com/). This program is very small yet very powerful. Irfanview will perform all the basics, cropping, downsampling, etc. The key feature is the enhance color options.

## Color Enhancing Options

Under the enhance color options you can adjust the color balance, brightness, contrast, gamma correction, and saturation. This will allow you to fix any imbalances in your photos caused by the lighting you use. Without a long explanation, the colors we see are directly affected by the light that shines on them, so if you use fluorescent light, tungsten (incandescent), or candle light you will get very different shades of color; this is a way to correct the colors in your photo. This area requires experimentation and deciding what looks best to you.

Irfanview doesn't stop there, it has many additions that can be installed allowing you to not only alter the image by converting it to black and white or sepia tone, but gives you a wide variety of photo alterations. Play with your pictures and see what you can create. Just to remember to be polite when you post your image, much of the world is still on dial up modems and it can take them some time to download your 4000 x 5500 pixel image. Keep your large original image for printing purposes, but share a smaller file that makes it easy for all to enjoy.

Now that you know how to optimize your photography, show us your customs!

# DIY Light Tent:
# Take Lighting to the Next Level!

Building a DIY light tent (also called a soft box) is very simple, inexpensive and will create a method to properly light your customs for photography. Most of the items needed to make a light tent are lying around your house. Here's a quick list:

**Materials**
Cardboard box
Construction Paper
White Bristol board
Wax Paper / Fabric
Light Source (Various colored bulbs)

**Tools**
Tape (scotch, double-sided, masking, painter's tape)
Glue / Glue stick
Ruler / Measuring Tape
Pen / Pencil / Marker
Scissors
Knife (X-acto)

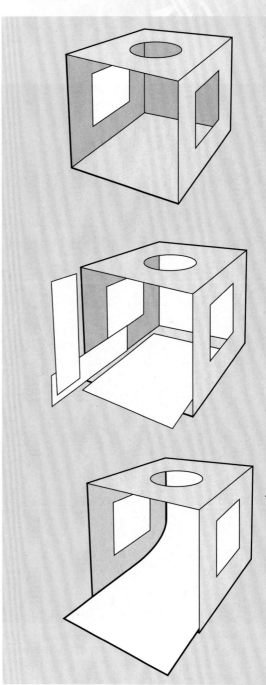

The cardboard box can be any size you want as long as you have enough light(s) to fill it. First, take the cardboard box and measure 2 inches in from each side using a marker on two sides: left and right. Connect the marks, which will result in a square (or rectangle, depending on box size) in the middle of the box. I cut out these regions on both sides of the box. Do not mark or cut the bottom of the box. As for the top, mark and cut it based on the size of the light you will be using. Once you have your all your marks connected, use your X-acto knife to cut these center sections out and remove them. Take your time and do your best to make the cuts as straight as possible. At this point you can remove the flaps of the box on one side; this makes the box open on one side.

Next line the inside of the box with white Bristol board (or white construction paper), this will reflect the light and is more durable than white paper. Using the measuring tape and marker, measure out two-inch strips on your white Bristol board; be sure that the board is long enough to fill the height and length of whatever box you have used. Since you cut two sides of your box you should need 8 two-inch wide strips, four for the inside of each side. As the opening in the top is custom to your light source being sure to replicate this for the Bristol board that will be covering the inside of the top. A glue stick or double-sided tape can be used to hold the Bristol board in place. Don't worry about overlapping strips or if the strips extend beyond your openings, it doesn't have to be perfect, but feel free to trim them to fit flush.

Now that the inside of the box has been lined in Bristol board (or construction paper) it is time to create the backdrop. Using Bristol board cut it to the inside width of the box, making sure it is much longer than the box. Place the long piece of Bristol board / construction paper into the box to where the piece curves and covers the entire bottom and the one side of the box that you did not cut a hole in. This is your backdrop. Avoid creasing as it will show up in your photos. You're looking for a nice smooth gradual curve here with plenty of room for the subject to sit/stand perfectly flat on the bottom of the box. Trim the excess board/paper that is sticking out the front of the box. Using painter's tape to hold the backdrop in place is a great option as it allows for its easy removal. This way different color backdrops can be used depending on the color of the subject. It can be very difficult to get good separation between a white clone trooper and a white backdrop.

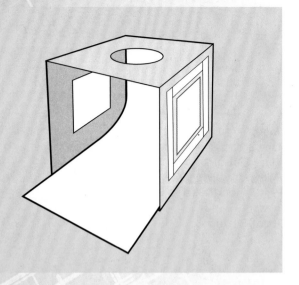

The final step in creating a light tent is to cover the side and top holes with light diffusing fabric. This can be nearly any type, white nylon, white fleece or white muslin fabric. Fabric is inexpensive at local fabric stores, but if you really want to pinch pennies there are a few options. One could use white under shirts, cut up the sides so you have two halves of the shirt to work with. If you use multiple shirts, be sure that the color matches or your end result may be uneven light diffusion inside your box. Ordinary wax paper could also be used. Measure your holes and the opening at the top of the box and cut your wax paper/fabric to be slightly larger than these measurements; I suggest at least half an inch of overlap all the way around. With your masking tape, tape all four sides down ensuring that the fabric/paper is tight and there is little or no slack. Any slack may cause odd shadows as the light will have a tough time diffusing uniformly through the material. Remember to NOT cover the hole facing your backdrop! This will need to be open so you can manipulate your subject and photograph through!

There you have it, a light tent!

*Note that Photographing black figures on a black background would be difficult. Not only can the background color be exchanged, but visual interest can be easily added with a decorative backdrop. Photograph by Michael "Xero" Marzilli.*

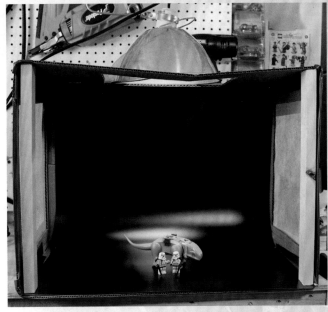

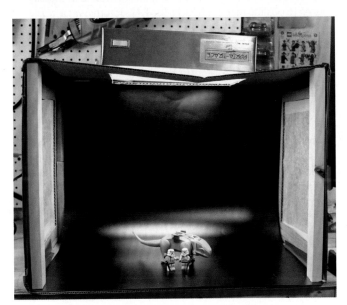

Note light tent construction, this particular tent is not lined due to its large size. Images correspond to the light effect images presented earlier. Photography by Michael "Xero" Marzilli.

### Yi Wong

**Major Area of Interest:** SciFi / Fantasy

**Years Customizing:** 5

**Web Gallery:**
*www.flickr.com/photos/23342881@N02/*

**What Customizer inspired you:** Arealight

**What got you into Customizing:** I got into customizing back in around 2007 when I had wanted to get the "Princess Storm" figure from an old Fantasy set, but decided that it wouldn't be that hard to make. And so I made my own version of the figure using spray paint, tape and felt-tip marker.

**Brief Customizing Background:** I started out customizing in 2007 and sporadically would make my own figures here and there. I eventually joined the KAM forum for a short time and eventually moved to posting mainly on Flickr. The KAM forums was actually where I had found out about Arealight, who's works had essentially helped inspire me to sculpt my own pieces. I eventually began talking to Hazel for a bit around 2010, where I actually began sculpting more. I used Milliput during this time. Eventually Arealight had asked me to be his spokesperson so I had accepted. And then in the summer of 2012 I started sculpting and casting my own parts; which I am still working on at this point.

**Favorite Customizing Tip or Trick:** Personally, I don't really have any super special tips or tricks; but I do really enjoy giving figures a colored metallic paint job. (I.E.: Cobalt or a metallic blue). This is actually pretty easy to achieve with a simple metallic paint and very thinned paint.

Basically, I first paint a base coat of black and follow up with another coat of silver. Afterwards I add multiple coats of thinned color paint.

Another useful tip for anyone who may start sculpting is that it is always much easier to take away than it is to add. With epoxy putties and modeling clays, I would suggest allowing the clay to harden a bit and then start adding your details with either a knife or some sort of sanding tool. And always smooth your work out with sanding or shaving (with a knife).

**Tools Needed for that Trick:** For paints, I personally like using the Citadel line of paints. They're pretty easy to use and usually won't take too much adjusting to get a color match to LEGO. They also have pre-thinned paints if you're worried about that; just look into glazes or washes. If that's not your thing, a simple wash is made by diluting the paint with either water or paint thinner. You'll have to be a bit more careful when using water though; if you add too much your paint will be far too diluted and will lose it's colors. As for sculpting, I just work in epoxy putties and use an X-acto knife and a sculpting pick.

## Dave "Geoshift" from Flickr

**Major Area of Interest:** SciFi, Military

**Years Customizing:** 4

**Web Gallery:**
*www.flickr.com/photos/41995251@N02/*

**What Customizer inspired you:** Jasbrick, pecovam, Pedro-79, eclipseGrafx (all from Flickr)

**What got you into Customizing:** Once I became a parent, I rediscovered LEGO and saw how the product had evolved over the years. I spent some time looking around Flickr and found the custom minifig groups. I was impressed by what people were doing in the hobby and wanted to get involved.

**Brief Customizing Background:** Many of my designs center around giving these little minifigs a rugged, "heavy" appearance. This is an interesting idea for me because it is contradictory to how they are commonly thought of. A common theme for me is creating minifigs in some sort of heavy armor, who appear to be rough, grizzled warriors. You won't find any smiling, cute little minifigs in my collection; these guys mean business! Again, this is not the way most people think of LEGO minifigs, which is one reason the style appeals to me.

**Favorite Customizing Tip or Trick:** All of my work involves painting pieces and parts, so my advice is to explore these options! A customizer can achieve some pretty amazing results with a little practice and patience.

**Tip #1:** Always prime your parts before painting and thin the paint slightly with water as you are working.

**Tip #2:** Workflow: Each painter will have their own methods, but a sample workflow is: primer, paint, detail, wash, highlight, varnish, done!

**Tip #3:** For added variety and customization options, use items from 3rd-party parts manufacturers in your minifig creations.

**Tools Needed for that Trick: Tip #1** and **#2:** Acrylic paints (not enamel), and detail brushes. I recommend brands like Citadel, Formula P3, and Vallejo, which are specifically designed for miniature painting.

**Tip #3:** Companies such as Brickarms, Brick Affliction, Tiny Tactical and eclipseBricks have created a multitude of quality items to be used in your customizing work, with new things appearing all the time. Check them out.

## The Clone Factory

**Major Area of Interest:** *Star Wars*, SciFi

**Years Customizing:** 9

**Web Gallery:**
*www.flickr.com/photos/theclonefactory/*

**What Customizer inspired you:** Recluce Mage, Fine Clonier

**What got you into Customizing:** After I watched *Star Wars: Revenge of the Sith* I thought about the unique LEGO figures that were not being produced. Commander Bacara had a very unique helmet and I thought that I could create his look in a LEGO style. This was before other customizers were able to produce a much cheaper and cleaner looking part using plastic injection.

**Brief Customizing Background:** My customizing style has evolved immensely over the previous 9 years. I first started with clay and modeled a unique part using a LEGO form. My Commander Bacara and Airborne Trooper helmets are good examples of items that were mostly created from clay. When LEGO started to come out with more unique parts I would use them by cutting and combining them to make a new unique item. The most recent example of this is the Clone Pilot helmet. In the last 2 years, I have started to use CAD design and rapid prototyping to create completely unique parts. The Z6 Rotary cannon and most of the body parts for Darth Maul were produced using this method. Through it all I constantly improved my decal and resin casting skills so that I was able to produce quality items.

**Favorite Customizing Tip or Trick:** When I cast items I use LEGO parts to create the set-up. My favorite set-up uses a LEGO cone with a round plate 1x1 to hold a helmet for casting. This allows for a very thin amount of casting silicone to run into the helmet that can easily be removed. This can also be achieved by filling a helmet with a plasticine but this can be difficult to clean after. The benefits are a cleaner and smoother mold.

## Kyle Swan

**Major Area of Interest:** Superheroes and Comics

**Years Customizing:** 2

**What Customizer inspired you:** Too many to name! In decals - eclipseGrafx, Christo, Levork, Tom Leech, Roaglaan and Kaminoan

**What got you into Customizing:** When I came out of my dark ages I started to take note of all the great minifigure customization happening online. I started a blog called 'The Ugly Duckling' which focused on LEGO minifigures and customization. I was able to meet, interview, and interact with a lot of great artists during the year or so the blog was active. Work and family life started to ramp up and I had to eventually put the blog to bed.

**Brief Customizing Background:** Wanting to still be active in the community, I started creating decals for customization on the open-source vector program, Inkscape. Since finding Inkscape and designing decals, I've largely been playing in Gotham City embarking on a series of Batman decals focused on his many iterations in comics over the years.

A lot of my early designs ended up having a lot of detail, incorporating highlights, textures, shadows. This appealed to my friend Andrew (Pecovam online) and he and I collaborated on a series of leg decals for his Halo and Gears of War customs. His company, Brick Affliction, has come out with some amazing ultra-detailed sculpts for minifigures. It has been a fantastic experience working with him.

In addition to my work with Andrew, I've also worked collaborations with Victor Fernandez (eclipseGrafx). I had designed a decal set for Bane from *Dark Knight Rises* and he thought my legs would be perfect for his printed minifigure 'Drifter'. He was able to get these designs printed and for sale in his store online and at conventions.

**Favorite Customizing Tip or Trick:** I can't stress how important a vector graphics program is for decal design. You don't have to spend hundreds of dollars on software either. Inkscape is a free, open-source vector program that I use for all my decal designs.

Aside from that I would also recommend finding as many reference materials as you can when planning a design. I look at reference images online. If there have been other printed toys that have focused on the character (Mighty Muggs, Minimates, Mezco, etc.), I'll look to see what design features they chose to emphasize and de-emphasize when finalizing my designs. Given the size of minifigures and the restrictions around what can and can't be seen at that size, you have to be careful in what to include.

**Tools Needed for that Trick:** You can download Inkscape at inkscape.org

I would also recommend opening an account on Pinterest (pinterest.com) to 'pin' images you come across when online. I can't tell you how many times I've seen an image when browsing online, only to never find it again when I needed.

## Mark Brockway
**Major Area of Interest:** SciFi

**Years Customizing:** 4

**Web Gallery:**
*www.flickr.com/photos/vieral/*

**What Customizer inspired you:** Fine Clonier, JasBrick, and Morgan190 stand out in my memory, though I am sure I am forgetting many others.

**What got you into Customizing:** When the *Star Wars* LEGO sets came out I started collecting LEGO again. I was thrilled with the partnership having been a long time *Star Wars* geek. The only challenge I found was not nearly enough focus on alien minifigures. So I decided to make what LEGO wouldn't: alien heads to fill in the gaps. I also did some research on the subject and found a whole new world …

**Brief Customizing Background:** I guess the beginning was in elementary school. I would make dragons out of Sculpey and sell them from my desk during recess and at lunch, which I had to close down because it "became a distraction." Later I started experimenting with WarHammer401K figures, branching out occasionally to other styles of lead figures. I learned a lot of painting and sculpting techniques painting lead figures. After high school I customized action figures for a while. Stopped customizing during college but I did a fair amount of sculpting with clay while in school. The *Star Wars* sets came along and I fell right into customizing the LEGO heads I wanted that LEGO didn't make.

**Favorite Customizing Tip or Trick:** Using epoxy putty over standard clay type products would be my best tip. It may be a personal bias, but epoxy putties like ProCreate allow a much higher level of detail. What I think of as clays like Sculpey, have only caused me frustration when customizing LEGO. I find epoxy putties much easier to manipulate before and after drying, whether by sculpting or Dremeling/sanding the piece to get a more polished look.

**Tools Needed for that Trick:** An epoxy putty similar to Procreate.

## MINGLES

**Major Area of Interest:** SciFi and Fantasy

**Years Customizing:** 3

**Web Gallery:**
*www.flickr.com/photos/60858662@N07*

**What Customizer inspired you:** ChocoBricks Customs

**What got you into Customizing:** I spent much of my childhood animation stop-motion animations with LEGO. As my YouTube account grew, I required the need for more realistic custom accessories and so the customization obsession began!

**Brief Customizing Background:** I first started customizing accessories when I became aware of the 3D printing service Shapeways. This allowed me to digitally design 3D CAD models of accessories which were scaled to fit LEGO Minifigures. Being able to physically create virtually any 3D model was the reason I stuck with it. I was now essentially able to make a LEGO version of anything I wanted. Due to 3D printing technology in its infancy, accessories have to be painted to look like the real thing. This however opens new doors to what material it can be created in. While the 3D printed plastics are obviously the closest thing to actual LEGO, the option to experiment with materials such as sterling silver and stainless steel also exists. This has allowed for some interesting results in both colors and texture.

**Favorite Customizing Tip or Trick:** When designing an item in 3D modeling software, I often try to make it look as 'LEGO' as possible. This basically includes bringing down the amount of detail and emphasizing the prominent features of the item. In terms of detail, we look at the item, for example a sword, and work out how it will look at about 45 millimeters. This allows us to disregard a large amount of the fine details and concentrate on the more distinct features such as the shape of both the hilt and the blade. These features are often thickened and rounded off to create a style similar to the LEGO Minifigure aesthetic.

**Tools Needed for that Trick:** Any 3D modeling software (e.g. Autodesk Inventor, Blender, SolidWorks, SketchUp) and an account with Shapeways.

## Nathaniel Ng Aka. Natsty

**Major Area of Interest:** Sci-fi, Weird War , Fantasy

**Years Customizing:** 3

**Web Gallery:**
*www.flickr.com/photos/-----natsty------/*

**What Customizer inspired you:** ~McLvin~ , Geoshift ,PEDRO-79 , Shobrick , pecovam, Jasbrick

**What got you into Customizing:** It all started 4 years ago. I was surfing the net for LEGO minifigures when I stumbled across some pretty cool figs by Jasbrick. The creativity in parts usage and colors transformed a ordinary minifigure into a piece of artwork. That got me instantly into customizing minifigs.

**Brief Customizing Background:** Customizing minifigs has always been a great hobby to me. Although much time and effort has been put into each figure, the outcome and feedback by others, especially my parents, has always been rewarding.

Every time I customize a minifig I learn something from it hence this allows me to improve over time.

**Favorite Customizing Tip or Trick: 1.** Pick a topic or figure with a level of difficulty reflective of your available time and painting ability.

**2.** Plan Plan Plan! Always plan what color and where are you are going to paint it.

## NickGreat

**Major Area of Interest:** Medieval Japanese, Anime

**Years Customizing:** 7

**Web Gallery:**
www.mocpages.com/home.php/3652

**What Customizer inspired you:** Jared Burks, Isaac Yue

**What got you into Customizing:** Seeing quality customs that actually look like LEGO.

**Brief Customizing Background:** It started with an entry of the Red Samurai, for Classic Castle contest in 2005. Although it did not win, my interest for minifig customizing continued. Was an active member in the MCN forum. Favorite pieces that I have done are my four Samurai and Dragonball series.

**Favorite Customizing Tip or Trick:** Cut and shape. I like to find ready LEGO pieces that require the least modification to acquire desired end result.

**Tools Needed for that Trick:** To ensure the desired piece is LEGO smooth after shaping, I use 1500 grit sandpaper followed by Brasso polish. Dremel with grinding disc (rough cut); Low grit sandpaper followed by 1500 grit; Brasso polish.

## Silentmaster 005

**Major Area of Interest:** Superheroes, Pop Culture, Movies/TV, *Star Wars*

**Years Customizing:** 5

**Web Gallery:**
www.flickr.com/photos/42306553@N06/

**What Customizer inspired you:** Fineclonier

**What got you into Customizing:** After coming into contact with Fineclonier.com, and seeing the decals he had, I decided to go out of the box. I wanted to make LEGO Figures that I knew LEGO would not create. Then after putting my custom figures on Flickr and seeing a whole community based on customizing, I just wanted to create more and more.

**Brief Customizing Background:** Fineclonier inspired me into making custom *Star Wars* characters that I always wanted, like Dash Rendar and a "Shadows of the Empire" Luke with the vest. After that, I saw all the decal options he had and just added items from all over the customizing world (Brickarms, Brickforge, Brickwarriors, Amazing Armory, MMCB, and many more). I started to get praise over the figures I have done, but I didn't feel like they were my own since I was using Finecloniers decals. So I started mixing up decals, using paint, modding LEGO parts, and started to create my own versions of TV and Movie characters. It was then that I could start calling figures my own. I still use Fineclonier as a guide and I will never stop because he is the best at what he does.

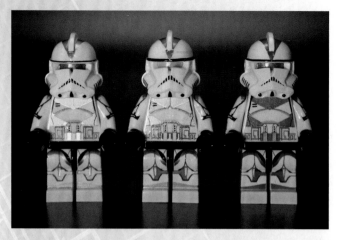

**Favorite Customizing Tip or Trick:** My favorite tip of the trade is painting. There are just some characters that need that extra paint to make them stand out. For example, if a figure needs to have short sleeves, you need either a decal or paint. I feel that the paint is the easiest solution.

**Tools Needed for that Trick:** I use Testors paint and the thinnest paint brush to apply the coats evenly. When making short sleeves on figures, I use the blue painters tape and mark off how far the sleeve should be.

### Michael "Xero_Fett" Marzilli

**Major Area of Interest (SciFi, History, Military, Etc.):** *Star Wars*

**Years Customizing:** 7+ years

**Web Gallery:**
*www.flickr.com/photos/54674651@N04/*

**What Customizer inspired you:** Fineclonier, Arealight, Christo

**What got you into Customizing:** Admiration for customizers' work and the desire to create custom figures of my own.

**Brief Customizing Background:** I specialize in water-slide decal application, custom painting and some parts modification.

**Favorite Customizing Tip or Trick:** Smooth decals applied to rounded areas (shoulders/arms, helmets).

**Tools Needed for that Trick:** X-acto knife, decal setting solution, decal softening solution, tweezers, Q-tip, patience and practice!

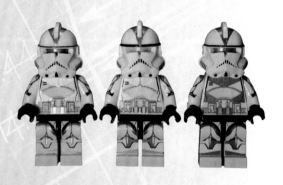

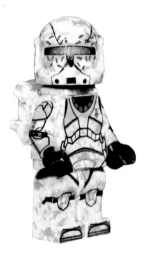